The Carbon Almanac

The Carbon Almanac

It's Not Too Late

Produced by
The Carbon Almanac Network
Edited by Seth Godin
Foreword by Seth Godin

PORTFOLIO | PENGUIN

Portfolio / Penguin

An imprint of Penguin Random House LLC

penguinrandomhouse.com

Most Portfolio books are available at a discount when purchased in quantity for sales promotions or corporate use. Special editions, which include personalized covers, excerpts, and corporate imprints, can be created when purchased in large quantities. For more information, please call (212) 572-2232 or email specialmarkets@penguinrandomhouse.com. Your local bookstore can also assist with discounted bulk purchases using the Penguin Random House corporate Business-to-Business program. For assistance in locating a participating retailer, email B2B@penguinrandomhouse.com.

Grateful acknowledgment is made for permission to reprint the following:

"A Brave and Startling Truth" from *A Brave and Startling Truth* by Maya Angelou, copyright © 1995 by Maya Angelou. Used by permission of Random House, an imprint and division of Penguin Random House LLC. All rights reserved.

"The Tyranny of Convenience" copyright © 2018 by Tim Wu. First published in *The New York Times* (Feb. 18, 2018). Reprinted with permission of Tim Wu.

Image credits: page 3 Peter von Cornelius via Getty Images; page 116 Leon Neal via Getty Images; page 211 Bettmann via Getty Images; page 274 photo by Anders Hellberg used via CC BY-SA 4.0 (creativecommons.org/licenses/by-sa/4.0/legalcode); page 284–85 "Driest Spot on Earth" copyright © 2012 by Lisa K. Blatt; comics by Dan Piraro used with permission of the artist; comics by Tom Toro used with permission of the artist; XKCD by Randall Munroe, xkcd.com; page 330–31 Shepard Fairey, © Obey Giant Art, Inc.

ISBN 9780593542514 (trade paperback)
ISBN 9780593542521 (ebook)

Printed in the United States of America
10 9 8 7 6 5 4 3 2 1

Table of Contents

Solutions   **157**

CHECK OUR WORK

The Almanac is based on thousands of sources. Don't take our word for it. Look for this number at the end of an article and then visit www.thecarbonalmanac.org/000 (but replace 000 with your article number). **Dig deep and share what you learn.**

www.thecarbonalmanac.org

Find the sources for all the quotations and fact boxes at ⊕ 888.

Foreword

This is a book about energy.

For more than a hundred years, we've had the opportunity to pump energy out of the ground virtually for free. We've used that cheap fuel to build the world around us, and we've created amazing things, wasted valuable resources, and made a mess while we were at it.

At the same time, this is a book about a different sort of energy. The energy of hope and connection. The ability that humans have to solve problems and to make things better.

It's not too late to make a difference.

But we're going to have to hurry. We can't waste a moment arguing about the size of our problem or mourning what used to be. Instead, we can lean into hope and connection.

The hope that comes from realizing that it's not too late.

And the power, nearly unlimited, that comes from coordinated action and community reinforcement. Connected we are far more effective than each of us acting individually.

This Almanac was created by more than 300 volunteers. Most of us had never met before committing to take coordinated action. Based in more than forty countries, from Benin to the Netherlands, from Australia to Singapore, we worked literally around the clock (time zones!) to build the book you have in your hands.

And if this book inspires you enough to share a copy with a friend, it will have been worth it. If it causes you and your friend to organize a circle of ten people, it will have made a difference. And if your ten people coordinate with ten other groups to cause organizational and cultural change to happen, it will be a success.

We live in an era of convenience, shortcuts, and sound bites. None of these are going to help us create a better tomorrow. Instead, we have the opportunity to focus on the things that really matter, and to do it with grace and urgency.

If not now, when?

Thank you for leading.

Seth Godin

The Carbon Almanac

Introduction

What is carbon, why does it matter, and why should I care?

The Four Horsemen of the Carbon Apocalypse

If we have the guts to focus on the big things now, the road forward will be far easier and more predictable.

The increase in carbon dioxide and other greenhouse gases in our atmosphere is threatening civilization. Throughout this Almanac, you'll encounter charts, graphs, and statistics that highlight and confirm the situation that we're in.

At the end of 2021, the global concentration of carbon dioxide in the atmosphere is more than 415 ppm, an increase of more than 25 percent in just 50 years. That growth is due to human activity. In order to reverse this trend and preserve the climate we all depend on, it needs to *decrease* dramatically in the next decade.

Four factors account for a significant portion of the carbon dioxide and other greenhouse gases released by humans. They are: **coal, combustion, cows, and concrete**.

Together **these four factors account for a projected 70 percent of our climate change problem**. And they are also among the greatest leverage points for reducing global emissions and achieving our carbon reduction goal.

All four of these factors are human choices, and all four have alternatives. Switching long-standing systems is never easy, but the path forward for each factor is straightforward (although each is going to be a challenge for society).

Coal

For hundreds of years, the primary function of coal has been to produce heat. Heat for shelter, for electricity, and for industry. The Industrial Revolution was made possible by coal because England had plenty of it, coal required little technology, and it was easy to burn.

Coal-fired steamships and trains then moved products across the globe. Centuries later, some countries around the world still depend on coal to either heat homes and food directly or to do so indirectly through coal-powered electricity plants.

All the coal on Earth was formed by the decomposition of ancient plants and animals buried hundreds of millions of years ago. Through this slow process molecules of carbon and hydrocarbons are compressed and become coal far beneath the Earth's surface.

As long as coal remains underground, the carbon is contained or "sequestered" and can safely be ignored. Extracting the coal through mining and then burning it releases the sequestered carbon and emits it into the atmosphere as carbon dioxide.

We're still burning coal: 14.8 gigatons of CO_2 worth in 2021. This represents **a quarter** of total global carbon dioxide equivalent emissions—the largest single source worldwide.

For the first time, there are plenty of cheap, reliable, low-polluting alternatives to coal. Now we need to put them into place.

Combustion

Combustion happens whenever we apply heat to fossil fuels in order to release their energy. Heat is produced, but stored carbon is also released.

Cars run on combustion, and so do gas-fired power plants and your backyard grill. (Coal combusts as well, but it's such a big part of the problem that it's worth treating as a separate category.)

Throughout this Almanac, you'll see examples of how people are reducing and replacing combustion with other smarter, more resilient sources of energy.

Cows

Compared to a coal plant, a cow seems benign. But cows produce methane, cows are near the top of the food chain, and cows live on a lot of land.

While there is far more carbon dioxide than methane released globally each year (70 times as much), methane has **84 times** the potential for trapping and heating the air over a 20-year period. The beef and dairy cattle industry is a primary source of greenhouse gas emissions globally.

Cows create methane through their digestive and waste elimination processes. Through the digestive process alone, each cow will belch 220 pounds (100 kilos) of methane every year.

The planet is home to an estimated 1.4 billion cows. In developing countries, meat consumption is increasing. In just 25 years, the consumption of beef in Asia is expected to triple.

In addition to the methane emissions from cows, the impact of cattle grazing affects climate change through the degradation of land and soil and the reduction of biodiversity. **The United States alone has 95 million cows and their grazing areas take up nearly half of the country's land.**

Reducing methane is a fast and powerful way to impact our climate.

Concrete

Concrete is everywhere. It's been around for centuries and is tough, versatile, and inexpensive. It's used for airports, buildings, bridges, dams, and roads. Following water, concrete is the most highly consumed product on the planet.

People are often surprised to discover that this one product contributes 8 percent of all global carbon dioxide emissions. In the last 40 years, we've tripled the per capita production of concrete, and as a result it's making an impact.

To create concrete, sand and gravel are added to cement, mixed with water, and poured into forms where it will dry and harden. No significant emissions are produced by this process. But first cement must be made, and that is where the carbon emissions come in. A very old and inefficient technology is used to make cement. But now there are new technologies available that produce far less impact.

China produces more cement than any other country, using more in the last three years than was used by the United States in all of the 20th century. India comes in second, followed by countries of the European Union.

These four factors—coal, combustion, cows and concrete—are responsible for a huge percentage of our challenge. And all of them will require systemic change—change that will happen if we act. Seeing the problem is the first challenge, and spreading the word to cause action is the next.

⊕ **003**

Change Is Here

Not everything can go up forever. For a hundred years or more, we've been cutting, burning, tilling, and trashing our way to the future.

And now there's no avoiding the truth of what's happened. Each of us is becoming aware that, like it or not, change is here.

But the change can be for the better. It can create jobs, enhance careers, and give us a chance to focus on what we've overlooked. We can reconsider how we spend our days, the way we treat each other, and how we create a better world.

The change is here. It's not too late. This is our moment to decide what to do with it.

🌐 009

803,719 BCE   0 CE   500   1000   1500

CO2 CONCENTRATION

400 ppm —

350 ppm —

300 ppm —

750 vehicle

500 vehicle

14 billion tons of CO₂

VEHICLES PER 1000 PEOPLE IN THE U.S.

ANNUAL CO2 EMISSIONS FROM COAL

7 billion tons of plastic

CUMULATIVE PLASTICS PRODUCTION

1900

1950

2000

The Tyranny of Convenience

" Convenience is the most under-estimated and least understood force in the world today. As a driver of human decisions, it may not offer the illicit thrill of Freud's unconscious sexual desires or the mathematical elegance of the economist's incentives. Convenience is boring. But boring is not the same thing as trivial.

In the developed nations of the 21st century, convenience—that is, more efficient and easier ways of doing personal tasks—has emerged as perhaps the most powerful force shaping our individual lives and our economies. This is particularly true in America, where, despite all the paeans to freedom and individuality, one sometimes wonders whether convenience is in fact the supreme value.

"Convenience decides everything"

As Evan Williams, a co-founder of Twitter, recently put it, "Convenience decides everything." Convenience seems to make our decisions for us, trumping what we like to imagine are our true preferences. (I prefer to brew my coffee, but Starbucks instant is so convenient I hardly ever do what I "prefer.") Easy is better, easiest is best.

Convenience has the ability to make other options unthinkable. Once you have used a washing machine, laundering clothes by hand seems irrational, even if it might be cheaper. After you have experienced streaming television, waiting to see a show at a prescribed hour seems silly, even a little undignified. To resist convenience—not to own a cellphone, not to use Google—has come to require a special kind of dedication that is often taken for eccentricity, if not fanaticism.

For all its influence as a shaper of individual decisions, the greater power of convenience may arise from decisions made in aggregate, where it is doing so much to structure the modern economy. Particularly in tech-related industries, the battle for convenience is the battle for industry dominance.

Americans say they prize competition, a proliferation of choices, the little guy. Yet our taste for convenience begets more convenience, through a combination of the economics of scale and the power of habit. The easier it is to use Amazon, the more powerful Amazon becomes—and thus the easier it becomes to use Amazon. Convenience and monopoly seem to be natural bedfellows.

Given the growth of convenience—as an ideal, as a value, as a way of life—it is worth asking what our fixation with it is doing to us and to our country. I don't want to suggest that convenience is a force for evil. Making things easier isn't wicked. On the contrary, it often opens up possibilities that once seemed too onerous to contemplate, and it typically makes life less arduous, especially for those most vulnerable to life's drudgeries.

But we err in presuming convenience is always good, for it has a complex relationship with other ideals that we hold dear. Though understood and promoted as an instrument of liberation, convenience has a dark side. With its promise of smooth, effortless efficiency, it threatens to erase the sort of struggles and challenges that help give meaning to life. Created to free us, it can become a constraint on what we are willing to do, and thus in a subtle way it can enslave us.

It would be perverse to embrace inconvenience as a general rule. But when we let convenience decide everything, we surrender too much.

Convenience as we now know it is a product of the late 19th and early 20th centuries, when labor-saving devices for the home were invented and marketed. Milestones include the invention of the first "convenience foods," such as canned pork and beans and Quaker Quick Oats; the first electric clothes-washing machines; cleaning products like Old Dutch scouring powder; and other marvels including the electric vacuum cleaner, instant cake mix and the microwave oven.

Convenience was the household version of another late-19th-century idea, industrial efficiency, and its accompanying "scientific management." It represented the adaptation of the ethos of the factory to domestic life.

However mundane it seems now, convenience, the great liberator of humankind from labor, was a utopian ideal. By saving time and eliminating drudgery, it would create the possibility of leisure. And with leisure would come the possibility of devoting time to learning, hobbies or whatever else might really matter to us. Convenience would make available to the general population the kind of freedom for self-cultivation once available only to the aristocracy. In this way convenience would also be the great leveler.

This idea—convenience as liberation—could be intoxicating. Its headiest depictions are in the science fiction and futurist imaginings of the mid-20th century. From serious magazines like *Popular Mechanics* and from goofy entertainments like "The Jetsons" we learned that life in the future would be perfectly convenient. Food would be prepared with the push of a button. Moving sidewalks would do away with the annoyance of walking. Clothes would clean themselves or perhaps self-destruct after a day's wearing. The end of the struggle for existence could at last be contemplated.

The dream of convenience is premised on the nightmare of physical work. But is physical work always a nightmare? Do we really want to be emancipated from all of it? Perhaps our humanity is sometimes expressed in inconvenient actions and time-consuming pursuits. Perhaps this is why, with every advance of convenience, there have always been those who resist it.

> **Perhaps our humanity is sometimes expressed in inconvenient actions and time-consuming pursuits. Perhaps this is why, with every advance of convenience, there have always been those who resist it.**

They resist out of stubbornness, yes (and because they have the luxury to do so), but also because they see a threat to their sense of who they are, to their feeling of control over things that matter to them.

By the late 1960s, the first convenience revolution had begun to sputter. The prospect of total convenience no longer seemed like society's greatest aspiration. Convenience meant conformity. The counterculture was about people's need to express themselves, to fulfill their individual potential, to live in harmony with nature rather than constantly seeking to overcome its nuisances. Playing the guitar was not convenient. Neither was growing one's own vegetables or fixing one's own motorcycle. But such things were seen to have value nevertheless—or rather, as a result. People were looking for individuality again.

Perhaps it was inevitable, then, that the second wave of convenience technologies—the period we are living in—would co-opt this ideal. It would conveniencize individuality.

You might date the beginning of this period to the advent of the Sony Walkman in 1979. With the Walkman we can see a subtle but fundamental shift in the ideology of convenience. If the first convenience revolution promised to make life and work easier for you, the second promised to make it easier to be you. The new technologies were catalysts of selfhood. They conferred efficiency on self-expression.

> ## If the first convenience revolution promised to make life and work easier for you, the second promised to make it easier to be you.

Consider the man of the early 1980s, strolling down the street with his Walkman and earphones. He is enclosed in an acoustic environment of his choosing. He is enjoying, out in public, the kind of self-expression he once could experience only in his private den. A new technology is making it easier for him to show who he is, if only to himself. He struts around the world, the star of his own movie.

So alluring is this vision that it has come to dominate our existence. Most of the powerful and important technologies created over the past few decades deliver convenience in the service of personalization and individuality. Think of the VCR, the playlist, the Facebook page, the Instagram account. This kind of convenience is no longer about saving physical labor—many of us don't do much of that anyway. It is about minimizing the mental resources, the mental exertion, required to choose among the options that express ourselves. Convenience is one-click, one-stop shopping, the seamless experience of "plug and play." The ideal is personal preference with no effort.

We are willing to pay a premium for convenience, of course—more than we often realize we are willing to pay. During the late 1990s, for example, technologies of music distribution like Napster made it possible to get music online at no cost, and lots of people availed themselves of the option. But though it remains easy

to get music free, no one really does it anymore. Why? Because the introduction of the iTunes store in 2003 made buying music even more convenient than illegally downloading it. Convenient beat out free.

As task after task becomes easier, the growing expectation of convenience exerts a pressure on everything else to be easy or get left behind. We are spoiled by immediacy and become annoyed by tasks that remain at the old level of effort and time. When you can skip the line and buy concert tickets on your phone, waiting in line to vote in an election is irritating. This is especially true for those who have never had to wait in lines (which may help explain the low rate at which young people vote).

The paradoxical truth I'm driving at is that today's technologies of individualization are technologies of mass individualization. Customization can be surprisingly homogenizing. Everyone, or nearly everyone, is on Facebook: It is the most convenient way to keep track of your friends and family, who in theory should represent what is unique about you and your life. Yet Facebook seems to make us all the same. Its format and conventions strip us of all but the most superficial expressions of individuality, such as which particular photo of a beach or mountain range we select as our background image.

> ## The paradoxical truth I'm driving at is that today's technologies of individualization are technologies of mass individualization. Customization can be surprisingly homogenizing.

I do not want to deny that making things easier can serve us in important ways, giving us many choices (of restaurants, taxi services, open-source encyclopedias) where we used to have only a few or none. But being a person is only partly about having and exercising choices. It is also about how we face up to situations that are thrust upon us, about overcoming worthy challenges and finishing difficult tasks—the struggles that help make us who we are. What happens to human experience when so many obstacles and impediments and requirements and preparations have been removed?

Today's cult of convenience fails to acknowledge that difficulty is a constitutive feature of human experience. Convenience is all destination and no journey.

Convenience is all destination and no journey.

But climbing a mountain is different from taking the tram to the top, even if you end up at the same place. We are becoming people who care mainly or only about outcomes. We are at risk of making most of our life experiences a series of trolley rides.

Convenience has to serve something greater than itself, lest it lead only to more convenience. In her 1963 classic, "The Feminine Mystique," Betty Friedan looked at what household technologies had done for women and concluded that they had just created more demands. "Even with all the new labor-saving appliances," she wrote, "the modern American housewife probably spends more time on housework than her grandmother." When things become easier, we can seek to fill our time with more "easy" tasks. At some point, life's defining struggle becomes the tyranny of tiny chores and petty decisions.

An unwelcome consequence of living in a world where everything is "easy" is that the only skill that matters is the ability to multitask. At the extreme, we don't actually do anything; we only arrange what will be done, which is a flimsy basis for a life.

We need to consciously embrace the inconvenient —not always, but more of the time. Nowadays individuality has come to reside in making at least some inconvenient choices. You need not churn your own butter or hunt your own meat, but if you want to be someone, you cannot allow convenience to be the value that transcends all others. Struggle is not always a problem. Sometimes struggle is a solution. It can be the solution to the question of who you are.

Embracing inconvenience may sound odd, but we already do it without thinking of it as such. As if to mask the issue, we give other names to our inconvenient choices: We call them hobbies, avocations, callings, passions. These are the noninstrumental activities that help to define us. They reward us with character because they involve an encounter with meaningful resistance—with nature's laws, with the limits of our own bodies—as in carving wood, melding raw ingredients, fixing a broken appliance, writing code, timing waves or facing the point when the runner's legs and lungs begin to rebel against him.

Such activities take time, but they also give us time back. They expose us to the risk of frustration and failure, but they also can teach us something about the world and our place in it.

So let's reflect on the tyranny of convenience, try more often to resist its stupefying power, and see what happens. We must never forget the joy of doing something slow and something difficult, the satisfaction of not doing what is easiest. The constellation of inconvenient choices may be all that stands between us and a life of total, efficient conformity.

... never forget the joy of doing something slow and something difficult, the satisfaction of not doing what is easiest. The constellation of inconvenient choices may be all that stands between us and a life of total, efficient conformity.

— Tim Wu, 2018

🌐 008

9

Understanding Carbon Lock-in

The current global economy relies heavily on the use of fossil fuels. The inexpensive and convenient power they produce, the assets invested, and the expectation of stability make fossil fuels the basis of the world's productivity.

Climate change offers a strong incentive to change the technologies we use to more climate-friendly ones. Yet governments all over the world have failed to implement policies to make the shift happen quickly. One reason for this is called *carbon lock-in*.

For 200 years, the world has industrialized. The steam engine enabled transport, which led to world trade, which led to increased demand, which created needs for insurance, investment markets, and more.

People depend on each layer of the economy for their livelihoods. Like a pyramid with a very large foundation, carbon has become the bedrock of most people's income.

In the last century, the world's population grew by billions of people. New jobs have been invented for most of these people. Technology has allowed them to be fed. Those jobs and that food have been based on an economy that has used carbon without regard for its real cost.

New needs arise as a society develops. Over the past century, human needs have developed from a secure food supply to the ability to travel, a steady energy supply, and recently, access to the internet. Competing technologies have emerged to satisfy these growing needs as well. In a free-market economy, the companies providing these technologies have an interest in gaining a dominant market share. Dominant technologies in turn create new industries. This begins a cycle of lock-in:

- Other technologies are ignored as system standards and investment returns make it difficult for them to enter the existing system.
- Specialized companies form to investigate subsystems of the technology and optimize tiny parts of the system.
- The number of existing players with something to lose is far greater than the number of innovators who believe they will benefit from change.
- Specialized education emerges to provide experts to run the established system in the future.
- Specialized knowledge is built to improve productivity and enable interoperability.
- Institutions emerge in order to regulate the technology, which are then run or heavily influenced by members of the dominant economic and social classes, who are dependent on the status quo.

Dominant technologies rely heavily on the system they are built into. They provide a network effect and become more valuable with the number of people using them. A car is only valuable if there is a road to drive on. An electric car is worthless if there is no infrastructure provided to charge it.

The incremental cost for new people to use the technology goes down after the initial supporting infrastructure is built. Existing infrastructure makes technologies more affordable, which leads to more users. More users create the need for even more infrastructure. The more the technology is used, the more people become attached to the value and benefits it provides. So the cycle continues.

This is how we end up in a world fueled by carbon, facing an existential threat in the form of climate change, but seemingly unable to alter the underlying systems at play. Sustainable technologies must overcome the initial hurdle of adoption and infrastructure in order to create the necessary change. Only then can they achieve lock-in to begin to reverse the underlying threat of climate change.

⊕ **006**

CO$_2$ & Dow Jones Industrial Average, 1960-2021

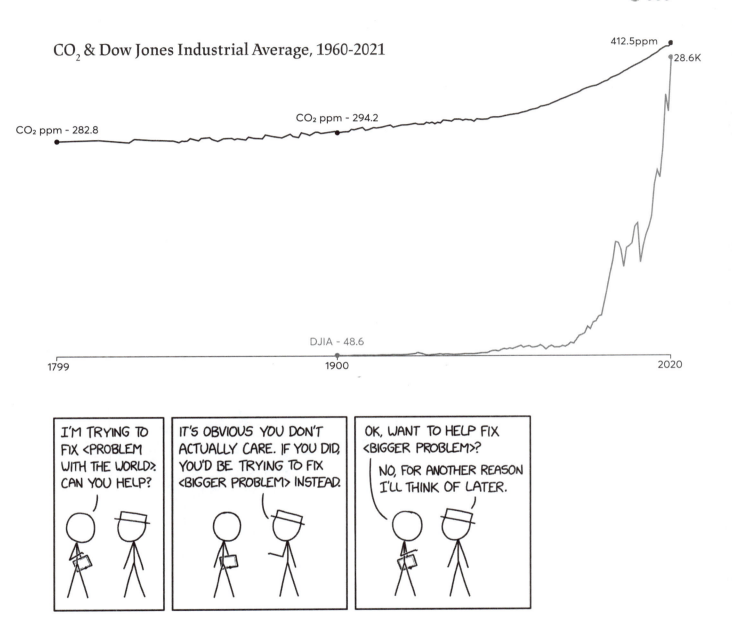

412.5ppm
28.6K

CO₂ ppm - 294.2

CO₂ ppm - 282.8

DJIA - 48.6

1799   1900   2020

The Wizard, the Prophet, and the Ostrich

Norman Borlaug won the Nobel Peace Prize for his work using technology to revolutionize farming. The Green Revolution he pioneered is estimated to have saved one billion people from starvation, overcoming a chorus of predictions that population growth would bring famine.

During those same decades, William Vogt kicked off the ecological movement, demonstrating that population growth was impacting the world we live in. He urged humanity to restrain colonization of the planet or face certain doom.

Charles C. Mann wrote about both men, describing Borlaug as *the Wizard* who believed that technology could ensure we build a healthier, more resilient planet, and Vogt as *the Prophet* who warned that growth inevitably brought doom. In many ways, these opposing views represent the two ways in which people view the challenges of climate change.

Some argue that only technological innovation and human progress offer hope for the planet. They argue for *more*—more power plants, more people, more technology. Others push for *less*. They seek dramatic reductions in the ways that humans interact with the natural world.

There's also a third group: the ostrich. When faced with uncertainty and fear, sticking their head in the sand is the natural response. They say there's a possibility that the climate isn't actually changing, or if it is, it's unrelated to human activity on the planet. Some even go so far as to claim it's a good thing for certain sectors of humanity.

Readers of this Almanac could find themselves adopting the viewpoint of the Wizard or the Prophet, sometimes both in the same afternoon. What's not possible, though, is to see the world through the eyes of the ostrich.

Let's begin by sharing the same reality.

🌐 002

Beyond the Polar Bear—Animals on the Edge of Extinction

There's no denying that adorable polar bears have mass appeal. But as the mascot for climate change, do they give us a false impression that climate change is happening "someplace else"? While the furry and easily anthropomorphized animals get all the attention, the problem is more widespread.

Climate change is here, right now, and eventually every living thing on the planet will be affected. About one million different species will be pushed into the danger zone in the coming decades and thousands are having a hard time adapting to human-provoked climate change right now.

Here's a tiny slice of the critters currently at risk:

Tigers		Giraffes	
Bumblebees		Insects	
Whales		Coral Reefs	
Asian Elephants		Ringed Seals	
African Elephants		Atlantic Cod	
Snow Leopards		Koalas	
Mountain Gorillas		Leatherback Sea Turtles	
Polar Bears		Adélie Penguins	
Monarch Butterflies		American Pikas	
Giant Pandas		Orange-spotted Filefish	
Delta Smelt		Green Sea Turtles	

Missing from this list are thousands of microorganisms, slugs, bugs, and other creatures that need a better PR firm.

🌐 **367**

Trash incinerators produce more than twice as many greenhouse gas emissions as coal-fired electricity plants.

Should You Opt In or Opt Out?

A quick online search for "email carbon" demonstrates that the internet thinks email is a significant contributor to greenhouse gases. There are dozens of articles and reports online (all seem to be based on an out-of-date estimate from 2010) that cutting back on email usage will have a significant effect on the climate.

It's tempting to believe that less email offers an easy answer because it seems like a fairly painless way to take action. Cut back on something you're doing too much of anyway and everyone wins.

It is also a useful example because we're confronted with choices like this every day.

Should we give up and realize we are each simply a drop in the ocean, unable to make a difference, or do we act and try to make an impact?

There are 300 billion emails sent every single day. Your absence on email, even if you're an evil spammer, is unlikely to be noticed or even missed.

But what if you used email to organize a thousand people to get an offshore wind farm approved? That single action would make a coal plant obsolete. That single coordinated activity aimed at changing the system would, by one estimate, remove six million metric tons of greenhouse gases a year.

When disconnected, each of us can accomplish very little. It won't matter a bit if we opt out. But we have far more leverage than we realize when we rally as a community to change systems.

🌐 001

We are the first generation to feel the effect of climate change and the last generation who can do something about it.

— Barack Obama

Game Theory

Game theory is the study of how people or organizations interact with each other in a situation where there are limited resources, desired outcomes, and a finite amount of time—which precisely describes the challenges of climate. What rules would have to be in place for countries to 'play a game' that would lead to a worldwide reduction in emissions? Why wouldn't wealthy, oil-rich countries cheat by free-riding when others are scaling back?

This is a version of the tragedy of the commons. If no one has an incentive to hold back, won't everyone graze their livestock until nothing is left?

Game theory tries to solve this challenge. The problem with reciprocity is that the countries that emit the most have the least need for reciprocal behavior by others as they are the wealthiest.

Climate degradation begins when someone dumps waste or burns fuel because it costs less than doing the resilient thing instead. Degradation can be avoided when all neighbors enjoy the same incentives. The three remedies are:
- Rewarding cooperation and reciprocity
- Limiting the temptation to free ride
- Punishing free-riders

If members of a group or different countries work together, systems can be built that lead to mutual rewards. When a marketplace is created where the invisible rules reward people for acting with the long-term in mind, that's what people and organizations are more likely to do. It turns out that social norms, pricing real costs into the system, and other interventions can change how organizations and countries behave.

Game theory therefore explains why some nations emit and avoid cleaning up—they get the benefits of cheap fuel while others pay for it with a changing climate and pollution.

Social norms have long changed the way organizations behave because they amplify beneficial long-term behaviors. The choices made by consumers and our responses to actions by producers can rewrite the rules that industries play by. Combined with fees and dividends related to carbon emission and capture, this can lead to a 'game' that the players win by cleaning up the mess that the last game created.

⊕ 004

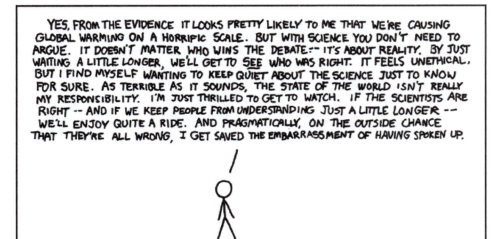

YES, FROM THE EVIDENCE IT LOOKS PRETTY LIKELY TO ME THAT WE'RE CAUSING GLOBAL WARMING ON A HORRIFIC SCALE. BUT WITH SCIENCE YOU DON'T NEED TO ARGUE. IT DOESN'T MATTER WHO WINS THE DEBATE — IT'S ABOUT REALITY. BY JUST WAITING A LITTLE LONGER, WE'LL GET TO SEE WHO WAS RIGHT. IT FEELS UNETHICAL, BUT I FIND MYSELF WANTING TO KEEP QUIET ABOUT THE SCIENCE JUST TO KNOW FOR SURE. AS TERRIBLE AS IT SOUNDS, THE STATE OF THE WORLD ISN'T REALLY MY RESPONSIBILITY. I'M JUST THRILLED TO GET TO WATCH. IF THE SCIENTISTS ARE RIGHT -- AND IF WE KEEP PEOPLE FROM UNDERSTANDING JUST A LITTLE LONGER -- WE'LL ENJOY QUITE A RIDE. AND PRAGMATICALLY, ON THE OUTSIDE CHANCE THAT THEY'RE ALL WRONG, I GET SAVED THE EMBARRASSMENT OF HAVING SPOKEN UP.

A Brave and Startling Truth

We, this people, on a small and lonely planet
Traveling through casual space
Past aloof stars, across the way of indifferent suns
To a destination where all signs tell us
It is possible and imperative that we learn
A brave and startling truth

And when we come to it
To the day of peacemaking
When we release our fingers
From fists of hostility
And allow the pure air to cool our palms

When we come to it
When the curtain falls on the minstrel show of hate
And faces sooted with scorn are scrubbed clean
When battlefields and coliseum
No longer rake our unique and particular sons
and daughters
Up with the bruised and bloody grass
To lie in identical plots in foreign soil

When the rapacious storming of the churches
The screaming racket in the temples have ceased
When the pennants are waving gaily
When the banners of the world tremble
Stoutly in the good, clean breeze

When we come to it
When we let the rifles fall from our shoulders
And children dress their dolls in flags of truce
When land mines of death have been removed
And the aged can walk into evenings of peace
When religious ritual is not perfumed
By the incense of burning flesh
And childhood dreams are not kicked awake
By nightmares of abuse

When we come to it
Then we will confess that not the Pyramids
With their stones set in mysterious perfection
Nor the Gardens of Babylon
Hanging as eternal beauty
In our collective memory
Not the Grand Canyon
Kindled into delicious color
By Western sunsets

Nor the Danube, flowing its blue soul into Europe
Not the sacred peak of Mount Fuji
Stretching to the Rising Sun
Neither Father Amazon nor Mother Mississippi who,
without favor,
Nurture all creatures in the depths and on the shores
These are not the only wonders of the world

When we come to it
We, this people, on this minuscule and kithless globe
Who reach daily for the bomb, the blade and the dagger
Yet who petition in the dark for tokens of peace
We, this people on this mote of matter
In whose mouths abide cankerous words
Which challenge our very existence
Yet out of those same mouths
Come songs of such exquisite sweetness
That the heart falters in its labor
And the body is quieted into awe

We, this people, on this small and drifting planet
Whose hands can strike with such abandon
That in a twinkling, life is sapped from the living
Yet those same hands can touch with such healing,
irresistible tenderness
That the haughty neck is happy to bow
And the proud back is glad to bend
Out of such chaos, of such contradiction
We learn that we are neither devils nor divines

When we come to it
We, this people, on this wayward, floating body
Created on this earth, of this earth
Have the power to fashion for this earth
A climate where every man and every woman
Can live freely without sanctimonious piety
Without crippling fear

When we come to it
We must confess that we are the possible
We are the miraculous, the true wonder of this world
That is when, and only when
We come to it.

— **Maya Angelou**
Monday, January 23, 2012

Climate Change for Rookies

What's all this talk about carbon?

What Is Climate Change?

Humans are causing the change in climate

The Earth's climate has fluctuated from the hot Jurassic period to the cold Ice Age. Since the Industrial Revolution nearly 140 years ago, the Earth's temperature has spiked, and scientists collectively agree that coal, oil, and gas burned by humans is the primary cause, followed by deforestation and intensive farming.

Fossil fuels

Coal, oil (petroleum), and natural gas are considered fossil fuels because like fossils, they formed deep in the Earth from the remains of plants, animals, and other living things from long ago. Coal and natural gas are burned at large power plants to generate electricity, and petroleum is the main ingredient in gasoline.

The greenhouse effect

Burning coal, oil, or natural gas releases carbon that mixes with oxygen to make carbon dioxide. Dubbed "the greenhouse effect," carbon dioxide and other gases act as a metaphorical roof of a glass greenhouse—letting sunlight pass through without allowing heat to escape. Other greenhouse gases that trap heat include methane and water vapor.

Until recently, some of the heat the sun provides has been able to easily escape the Earth's atmosphere, thus keeping the Earth's temperatures constant. Now, the build-up of greenhouse gases is insulating the Earth like a blanket and causing temperatures to suddenly climb.

The impacts of 1°C

Humans emit carbon simply by charging phones, making cookies, and driving to the store. As a result, the Earth has abruptly warmed about 1°C/1.8°F because we're burning fossil fuels for daily activities.

1°C/1.8°F may not seem like much of a temperature increase, but like a fever, it's enough to destabilize the Earth and cause extreme weather, including:

· Hurricanes
· Snowstorms
· Heatwaves
· Downpours
· High winds
· Droughts
· Flooding
· Landslides
· Wilder winter weather

SEVERITY OF WEATHER

We are called to be architects of the future, not its victims. [The challenge is] to make the world work for 100% of humanity in the shortest possible time, with spontaneous cooperation and without ecological damage or disadvantage of anyone.

— R. Buckminster Fuller

Because 1°C/1.8°F is an average global temperature increase, the rise in temperature for a particular area is often considerably different. For example, average temperatures have risen by 1.5-2°C over the Arctic.

While 1-2°C is a significant temperature change for the Earth, individuals might not feel the rise in temperature on any given day but instead recall a week of record-breaking heat or rainfall. Even a chubby squirrel hints at temperature increases, since more than enough food is available without the usual snow cover.

Ten years to act

2020 was the hottest year on record, and scientists report humanity has about ten years to drastically cut its carbon emissions before damage to the Earth is irreversible.

Climate change is complex, and there are no magic bullets or easy solutions. On one hand, while banning carbon-emitting concrete might seem plausible, developing nations depend on inexpensive concrete to build affordable buildings. Climate change solutions include embracing solar and wind–generated electricity to end our dependence on gasoline and oil, as well as changing how we eat and travel.

While we're working on reducing carbon, the world will still need to heat buildings, drive vehicles, and charge laptops, so individual efforts to stop climate change are valuable but limited. However, electing candidates committed to climate change initiatives and policies is considered one of the most effective ways to reduce emissions on a large scale.

Since the Industrial Revolution nearly 140 years ago, the Earth's temperature has spiked and scientists collectively agree that coal, oil, and gas burned by humans is the primary cause, followed by deforestation and intensive farming.

MAKE AN IMPACT

The Imperial College of London ranked nine things people can do to make an impact on climate. The first is the most important by far, and the ninth is why this Almanac exists...

1. Make your voice heard by those in power
2. Eat less meat and dairy
3. Cut back on flying
4. Leave the car at home
5. Reduce your energy use (and bills)
6. Respect and protect green spaces
7. Invest your money responsibly
8. Cut consumption and waste
9. Talk about the changes you make

🌐 **354**

The Greenhouse Effect

FAST FACTS ON THE GREENHOUSE EFFECT

Because CO_2 makes up about 80 percent of all greenhouse gases and is the most significant contributor to human-caused climate change, seeing or hearing the word "carbon" usually implies *all* greenhouse gases.

The primary greenhouse gases include:
   Carbon dioxide (CO_2)
   Methane (CH_4)
   Nitrous oxide (N_2O)
   Fluorinated gases
   Water vapor

Burning fossil fuels adds greenhouse gases to the atmosphere, traps heat, and causes the Earth to warm.

Greenhouse gases are released when humans burn fossil fuels like oil, natural gas, and coal to provide energy for daily activities. These gases rise into the atmosphere and insulate the Earth, causing temperatures to rise.

It's almost like the glass of a greenhouse roof. Carbon dioxide and other gases allow sunlight to pass through to the Earth but hold in heat. This happens because the incoming sunlight is reflected back by the Earth as infrared radiation, which can't easily escape back to outer space because greenhouse gases are present.

The 1°C/2°F increase in temperature that's occurred in the past century is a bit like that infant with a fever. A small change makes a huge difference. This temperature rise has destabilized the Earth and is causing severe weather like hurricanes, heavy downpours, flooding, droughts, and even snowstorms.

⊕ **753**

GREENHOUSE GASES

What's All This Talk About Carbon?

Anytime something is plugged in, made in a factory, or driven from here to there, carbon is released.

Carbon is present in all living things, but it's become problematic in the last 150 years as humans have innovated and the world has become industrialized.

The discovery of Earth's abundance of coal—which is primarily carbon—is considered one of the most significant influences leading to the Industrial Revolution, as coal could be used to fuel steam engines in trains, ships, and machinery.

As humans innovated, more and more carbon was released by burning coal, oil, and gas to fuel vehicles, generate electricity, and run machinery.

Here's the problem: When carbon mixes with oxygen, it forms carbon dioxide (CO_2), which traps heat above the Earth and causes temperatures to rise.

We're already feeling the physical and political effects of the changes in temperature over the last century. Our infrastructure is on the verge of being overwhelmed.

⊕ **751**

Weather vs. Climate

Weather is not the same as climate, but they're certainly related. Think of them like cousins.

Weather refers to the day-to-day atmospheric conditions, like an overnight snowstorm or a sunny afternoon.

Climate refers to the *overall* weather of a region, like the typical conditions you'd expect in Aruba during February.

Because of climate change, the expected weather for an area is often no longer the weather residents experience. Freezing temperatures in Texas or droughts and flooding in California indicate that residents can't count on "normal" weather conditions anymore.

⊕ **752**

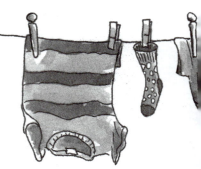

Invisible Carbon Emissions

Carbon dioxide is colorless and essentially invisible to the naked eye. Clothes tumbling in the dryer don't produce "visible emissions" that can be seen, but carbon-emitting coal or other fossil fuels are burned at faraway power plants to generate the electricity powering the appliance.

Understanding good advice:

Use a clothesline.
Clothes dryers rank as one of the highest carbon-emitting appliances in homes. Household appliances are powered by carbon-emitting fossil fuels used to generate electricity.

Use a refillable water bottle.
Heavy water bottles are transported to market by carbon-emitting trucks, and plastic bottles are made from carbon-emitting fossil fuels. Plus, only 9 percent of plastic is ever recycled.

Turn off the lights / heat / TV / air conditioning.
Burning coal or oil generates the power for roughly 70 percent of anything that's plugged in.

Eat local produce.
Local goods travel shorter distances and require fewer carbon-emitting trucks or planes to reach stores.

Put on a sweater.
Heating oils and natural gas are fossil fuels that emit carbon when burned.

Bring your own grocery bags.
Disposable plastic bags are rarely recycled, take hundreds of years to break down in landfills, and leach toxins into the soil as they decompose. Plastic is also made from carbon-emitting fossil fuels.

Use cold water for laundry. Take shorter showers. Turn your water heater down to 120°F.
The fuel required to heat water releases carbon; keeping water hot requires constant carbon-emitting energy.

Try video conferencing instead of plane travel.
A one-way plane trip from San Francisco to London emits twice as much carbon as a family car driven for a year.

Drive an electric car.
Unlike gasoline-powered cars, driving an electric vehicle produces no tailpipe emissions. However, if the power plant burns fossil fuels to generate electricity to charge that vehicle, some carbon pollution is still created, though far less than gasoline vehicles. As more companies shift to generating solar and wind power, the emissions to charge electric vehicles will decrease or disappear entirely.

🌐 **750**

How Much Is a Metric Ton (Tonne)?

It's hard to imagine an invisible gas having much weight at all, let alone it weighing a ton.

A metric ton is the unit of mass that scientists use to measure the weight of carbon dioxide.

One Metric Ton = 1,000kg (or 2,204.6 pounds) and is about the same as the weight of about 440 bricks or a great white shark.

US SHORT TON

Not the same thing but close...
A customary US ton weighs 2,000 pounds, so it is close but not identical to a metric ton.

Imagining carbon emissions

One metric ton of carbon dioxide takes up about as much space as a 10-meter cube—a cube whose sides would equal the length of a telephone pole.

In New York City, nearly two metric tons of carbon are emitted *every second*, mainly by buildings that burn fossil fuels to generate electricity. That's about 150,000 metric tons of carbon *a day*.

In the United States and Canada, each citizen averages a little over 14 metric tons of carbon emissions per year (the weight of about 6,300 bricks). To reach our goal of net–zero emissions by 2050, carbon emissions in the atmosphere need to end up at about one metric ton per person (or approximately 440 bricks).

Cutting emissions from 6,300 to 440 bricks per person will require individuals to:

1. Eliminate luxuries like airplane travel, plastics, air conditioning, and eating meat.
2. Elect officials to enact sweeping national change, shifting our energy sources from carbon-polluting fossil fuels to renewables like solar and wind.
3. Pressure large businesses to do business in a climate-friendly way.
4. Invest in carbon removal technologies and reforestation to balance the carbon still emitted.

Citizens of China emit only about one-third as much carbon as Americans, but China leads the world in emissions since China's population is so large. In smaller countries like Mozambique, each citizen already averages less than one metric ton of carbon per year.

Note: Technology that removes carbon from the atmosphere has yet to be perfected, so it's impossible to predict precisely how much carbon emissions each individual will actually have to cut to reach net-zero emissions.

🌐 **754**

Current US emissions in tons per person

2050 goal of per-person emissions

Fast Facts and Definitions

These at-a-glance definitions offer a quick understanding of common climate change terminology:

Carbon dioxide: Carbon that's released by burning oil, gas, and coal combines with oxygen and becomes carbon dioxide (CO_2), the primary driver of climate change.

Carbon emissions: While *all* greenhouse gases are collectively referred to as "carbon," there are significant differences in impact. Some greenhouse gases are hundreds of times more potent than carbon dioxide.

dead micro-organisms

400 million years ago

anaerobic decay under sand and silt and mud

100 million years ago

PRESSURE

HEAT

CRUDE OIL

Climate change: The shift in the Earth's temperature and ecosystems, including variations in precipitation, sea level, and agriculture.

Coal: A non-renewable fossil fuel usually burned to generate electricity at power plants for transmission to homes and businesses.

Emissions: Greenhouse gases released into the air. Can be created by burning fossil fuels or other human activities.

Fossil fuels: Coal, oil, and natural gas that formed deep within the Earth, millions of years ago from the remains of living organisms.

Global warming: An increase in the average surface temperature of the earth caused by rising greenhouse gases.

Greenhouse gases: Mostly carbon dioxide, methane, and water vapor that insulate the Earth and cause temperatures to rise.

Methane: The second most common greenhouse gas. It is 84 times more heat retaining than carbon dioxide over a 20 year period. Methane is released by cows during digestion, when industry burns natural gas, and when landfills decompose.

Mitigation: An effort that decreases or prevents the release of greenhouse gases, including planting carbon-absorbing trees and using renewable energy.

Natural gas: A non-renewable fossil fuel used mainly to heat buildings and generate electricity.

Non-renewable energy: Energy derived from natural carbon-emitting fuels such as oil, natural gas, and coal that will eventually run out.

Oil or Crude oil: A non-renewable fossil fuel. It is converted to gasoline, diesel, and also heating oil. It can be burned to generate electricity. Plastics are usually made from oil as well.

Renewable energy: Naturally-replenished and non–carbon–emitting energy like sunlight, wind, waves, and geothermal heat located deep within the Earth.

Sea-level rise: A phenomenon caused by rising temperatures, melting glaciers, and the expansion of water.

🌐 **756**

YOU CAN MAKE A DIFFERENCE

Visit **www.thecarbonalmanac.org** and sign up for **The Daily Difference**, a free email that will connect you with our community. Every day, you will join thousands of other people connecting around specific actions and issues that will add up to a significant impact.

CHECK OUR WORK

The Almanac is based on thousands of sources. Don't take our word for it. Look for this number at the end of an article and then visit www.thecarbonalmanac.org/999 (but replace 999 with your article number). **Dig deep and share what you learn.**

www.thecarbonalmanac.org

Climate Change in Front of Your Eyes

Climate change isn't happening somewhere "over there." It's happening right here in our lives and neighborhoods. Some effects of climate change that are right on our doorstep:

At home

Blackouts
Flooded basements
Internet service disruptions
Cell phone service outages
Frozen gutters
Fallen trees
Higher taxes
Unemployment
Higher electric bills
Soaring grocery bills
Skyrocketing insurance costs
Declining home values
Uninsurable homes

Around town

Potholes
Traffic
School closings
Melted power lines
Downed power lines
Flooded subways
Sewage backups
Detours
Contaminated water
Dam failures
Cracked pavement
Lower reservoir levels
Collapsed bridges
Street flooding

Health

Food-borne illness
Heatstroke
Hypothermia
Asthma
Hay fever
Lyme disease
Food insecurity

Recreation & Travel

Soggy golf greens
Less snowpack
Fewer ski days
Canceled events
Airplane turbulence
Travel delays
Red tide algae blooms
Beach erosion
Loss of tourism revenue

Outdoors

Pollution and smog
Changes in growing seasons
Declining crop yields
Increased pollen levels
Watering restrictions
Rising populations of mold,
    disease-carrying mosquitoes,
    and invasive plants
Later fall foliage
Earlier maple tapping
Trees and plants flower sooner
 and produce less fruit
Chubby squirrels
Fewer butterflies
Bears shortening hibernation
Widely spaced tree rings
Avalanches
Shellfish destruction

Weather

Wildfires
  Drought
Flooding
  Tidal surges
Severe storms
 Heavy downpours
  High winds
Tornadoes
  Hurricanes
  Snow in typically warm areas
Concurrent disasters
Consecutive disasters
  Heatwaves

🌐 079

What Is Net Zero?

Imagine a scale with carbon polluters like power plants and gas-powered cars on the right, and carbon absorbers like trees and oceans on the left. When the scale balances, emissions are net zero.

Achieving net-zero emissions by 2050 is a goal. However, to reverse the existing damage, we'll have to go beyond net zero and produce *fewer* greenhouse gases than the Earth can absorb.

It's impossible to banish *all* carbon emissions; in order to reach net zero, emissions need to be reduced to the point where natural carbon sinks like trees and innovative technologies can remove the same amount of carbon emitted.

Achieving net-zero emissions requires both abandoning fossil fuels and investing in innovation to remove carbon.

Nobody knows exactly what life will look at in 2050, but based on technologies currently in the works now, this is one scenario:

A day in the life of net-zero emissions

8 AM Toss off a light blanket and step into a 21°C/70°F room. The well-insulated house has been retrofitted with triple-paned windows that keep the smart home at a steady 21°C/70°F year-round using an electric heat pump which also cools the house. Everything plugged in uses electricity generated from the neighborhood solar panels or purchased from solar and wind power plants.

9 AM The vintage 2019 coffee maker that originally used plastic coffee pods was upgraded two decades ago to use compostable bioplastic coffee pods. Almond and soy milk are the main ingredients in lattes now, but some still indulge in cow milk during the holidays.

10 AM Commuting to work only occurs a few days a month using electric trains and shared ride services.

11 AM The neighborhood goats arrive for their weekly feast. A herd of goats is enough to keep the community nicely pruned and free of weeds. The grass is long gone, but the plant seeds that were grown a few years ago were developed to need no additional watering and to grow reasonably short.

NOON A casual lunch with friends includes a plant-based burger. Animal meat is expensive and not widely available, so it's eaten only for special occasions. Instead of a furnace, the restaurant uses a geothermal heat pump that captures heat inside the Earth to keep the temperature toasty inside the restaurant when it's cold outside. There are also awnings, automatic shades, and large shade trees to help regulate the temperature inside.

1 PM Traveling back home to finish the rest of the workday, passing fields that used to be home to cattle and dairy farms and that now contain carbon scrubbers. These help the trees and oceans reduce the carbon in the air.

2 PM Video call with friends to hear about their 25th-anniversary trip overseas. Plane travel still occurs, but since there are expensive carbon removal surcharges they are reserved for special occasions.

4 PM Make a yearly service appointment for the whole-house battery—programmed to pull electricity from the grid when it's cheapest and ensure stored energy is readily available for the home (even when the sun is down or there's no wind). In fact, to maximize solar input, roofs in the community are pitched in a checkerboard design to ensure no neighbor casts a shadow on another's home.

5 PM Sautéed eggplant and greenhouse-grown vegetables are cooked on an induction stove, which targets the heat to the food without a lot of energy waste.

9 PM Electric vehicle charging finishes in the garage. The system is optimized so that charging occurs when the grid has less demand. The garage has two bays because it was built in the 2030s when families still had multiple cars. Most homes built now have one bay only, and if more than one family member needs to get around simultaneously, homeowners use ride-sharing services.

10 PM Targeted bed environments turn on to save energy at night. These "smart" mattress pads target warmth or coolness directly to the person in bed and raise and lower the temperature all night for a deeper sleep.

⊕ **755**

⊕ **999**

Don't take our word for it

Each article in this Almanac is based on our review of thousands of different sources. Visit **www.thecarbonalmanac.org/999** (insert correct number) to see links, explanations and updates.

Dig deep and share what you learn.

10 Myths About Climate Change

Myths about climate change persist. Despite being completely debunked by climate scientists, they resurface regularly. According to the UN's Intergovernmental Panel on Climate Change, the scientific evidence for warming of the Earth's climate system is indisputable.

During the Industrial Revolution, people started burning coal and other fossil fuels to power factories, smelters, and steam engines. Burning fossil fuels adds greenhouse gases to the atmosphere, which has increased the average global temperature by 1°C/2.2°F since 1880.

Myth 1

Climate change is nothing new: the climate is always changing

Seventeen of the 18 warmest years in recorded history have occurred since 2001. Human activity—burning coal, oil, and gas—is fueling this change. The World Wildlife Fund reports that the rapid changes seen today would otherwise have occurred over hundreds of thousands of years, not mere decades.

Myth 2

The Earth is not warming; it's still cold outside!

When the ice in your glass melts, your drink gets colder—for a while. As the Earth warms, there will be less snow cover and less sea ice around the North and South Poles. These large areas of low–pressure cold air are referred to as a polar vortex—the counter-clockwise flow of air that helps keep colder air near the Poles. When these vortexes are destabilized by warmer air, the result is cold snaps and freezing temperatures in areas that are normally warm. This happened to Texas in 2020. This destabilization can also cause an increase in atmospheric moisture, resulting in heavier rain, hurricanes, and snowstorms.

Myth 3

Renewable energy is expensive

The price of solar power and onshore wind has plummeted in the last decade, making solar and wind energy two of the most economical methods of generating electricity.

Global average surface temperature anomalies (°C)

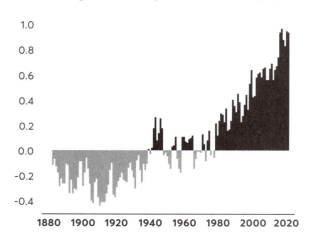

The price of electricity from new power plants (kWh)

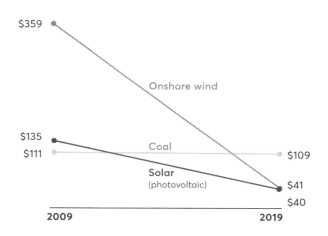

Myth 4

Solar and wind can't work because they are dependent on the weather

As advances are made in batteries and electricity storage, excess energy from sunny and windy days can be stored and then deployed later on cloudy, cool days. While there is not yet cost-efficient storage capacity for around-the-clock worldwide renewable energy, storage capacity is growing to meet demand.

Myth 5

Most people don't believe in climate change

In 2020, according to the Yale Center for Climate Communication, 55 percent of the United States population is concerned or alarmed by climate change; only 20 percent are dismissive or doubtful.

Myth 6

Recycling plastic helps address climate change

Regardless of the symbol stamped on plastic, only 9 percent of plastic is recycled, leaving the rest to be incinerated or to accumulate in landfills and oceans. The "chasing arrows" symbol and coordinating number embossed on plastic products is simply a way to identify the plastic resin used to make the product. When plastic or other disposed containers are burned, more carbon dioxide and other greenhouse gases are released.

9%

Regardless of the symbol stamped on plastic, only 9 percent of plastic is recycled, leaving the rest to be incinerated or accumulated in landfills & oceans.

Myth 7

A #6 Styrofoam cup or takeout container placed in recycling bins helps the environment

Since polystyrene is mostly air and only a tiny fraction of the plastic is recoverable, most communities cannot economically recycle #6 plastics. As a result, single-use plastics get replaced by more single-use plastics.

Myth 8

Depleting ozone is the principal cause of climate change

Ozone loss is not the cause of climate change. According to NASA, while the ozone depletion that has already occurred in recent years is not helping climate change, its contribution is small compared to all other heat-trapping gases.

Myth 9

Climate change does not affect me personally

Because climate change is happening gradually, it tends not to command the same attention as an abrupt change would. In fact, 85 percent of the world's population has already experienced climate change or its effects firsthand, usually via severe storms, blackouts, heatwaves, and droughts. Regardless of whether your home is flooded today, the supply chain, the economy, and your livelihood all depend on people outside your neighborhood.

Myth 10

It's too late—nothing can be done about it

It's not too late. According to the United Nations, the Earth has about ten years to cut greenhouse gas and carbon emissions to prevent irreversible damage to the climate. Many organizations are working on cutting emissions to ensure a future for humans on Earth.

🌐 **342**

20 Truths About Climate Change

1. 99.5 percent of climate scientists agree that humans are causing climate change.

2. Greenhouse gases, like methane and carbon dioxide, behave much like the glass of a garden greenhouse. Sunlight can pass through the atmosphere, but heat is prevented from leaving. This causes the Earth's temperatures to rise.

3. Extreme weather like flooding, heatwaves, snow, downpours, and droughts is amplified by the warming of the atmosphere.

4. The ozone layer is high overhead and protects the Earth. That's not the same as ozone that's caused by pollution, which is a greenhouse gas that humans emit into the environment.

5. The Earth's climate has always changed. However, it's now happening over decades instead of hundreds of thousands of years.

6. The amount of CO_2 in the atmosphere is higher than it's been in 2 million years.

7. A global rise in temperature can cause drops in temperature on a local level. That's why snow in normally warm areas like Texas is a symptom of the atmosphere heating, rather than a refutation of it.

8. Ice reflects sunlight and keeps the Earth cool, so when ice sheets and glaciers melt, the water they create absorbs the sunlight's heat and oceans warm up even faster.

9. A warmer atmosphere can hold on to more moisture, which causes heavier and more frequent rainfall.

10. The last seven years have been the warmest seven years on record.

11. Nine of the costliest mainland US hurricanes on record have occurred in the past 15 years.

12. Sea levels are rising, and the rate at which they're rising is getting faster. Even if the world made a radical switch to low greenhouse gas emissions right away, the increase is predicted to be at least 30 centimeters by 2100. The worst-case scenario—if we do nothing to stop rising temperatures—could cause sea level rise of 2.5 meters.

13. 634 million people live within 33 vertical feet of sea level and are at high risk of flooding and having their land become submerged.

14. Between 1982 and 2016, the snow season in the western United States decreased by 34 days.

15. Some insects are aggressively killing carbon-absorbing trees because recent winter temperatures aren't cold enough to control their spread.

16. Higher temperatures increase the spread of disease.

17. Between 150 and 200 species of plants and animals are now becoming extinct each day.

18. Of all the plastic ever made by mankind, only 9 percent has been recycled. 12 percent has been incinerated and 79 percent remains in landfills or as litter in the environment. Single-use plastic is replaced by new plastic, releasing more greenhouse gases.

19. Decomposing landfills emit methane, a greenhouse gas 84 times more potent than carbon dioxide.

20. During the 2020 Covid-19 pandemic, carbon emissions decreased by 5.8 percent, equivalent to the total carbon emissions of the European Union.

🌐 **032**

The Diffusion of Innovations

Climate change was first discussed by the scientific community more than fifty years ago, and Exxon scientists were clear about the impacts in the 1980s. Yet widespread acceptance of the facts about our climate is slow.

This isn't surprising.

Everett Rogers wrote about this phenomenon in 1962 when he formulated his "Diffusion of Innovations" theory. He described how ideas spread through a population. They never arrive all at once and are never embraced by everyone at the same time or in the same way.

Human beings are slow to change beliefs and actions, though some people are more eager to do so than others. Not in every area, not all the time, but in areas where they are open and eager to new ideas. This happens regardless of the efficacy of the intervention or the reliability of the facts a new idea is based on.

He proposed that for every sort of idea or area of interest, people fall into one of five categories:

1. **Innovators:** These are people who enjoy going first. They latch onto a new idea or innovation simply because it's new, not because it has to be proven to be correct or useful.

2. **Early Adopters:** These are people who, at this particular cultural moment, enjoy leadership roles and embrace change. They need to first see that the idea or technology is truly better. But because they enjoy leading, they're the most likely to evangelize a new idea to others.

3. **Early Majority:** These people aren't leading, but they choose to adopt new ideas before the average person. Following an early adopter gives them satisfaction and a measure of status.

4. **Late Majority:** Perhaps this group could be called skeptical, but when enough of the early majority shift gears, and when they've heard from early adopters with care and persistence, they're likely to come along.

5. **Laggards:** People who choose to be in this group on a given issue aren't going to change their minds easily, and in fact, may never do so. The best outcome may lie in ignoring them and focusing on the other groups instead. Averaging less than one-sixth of most populations, avoiding them is usually the most productive strategy.

🌐 **353**

INNOVATORS	EARLY ADOPTERS	EARLY MAJORITY	LATE MAJORITY	LAGGARDS
2.5%	13.5%	34%	34%	16%

Climate Change Actions from Large to Small

While everyone agrees that recycling is good, most people assume it's more effective than it is in reality.

Only 9 percent of all plastics are ever recycled, and that includes what's actively put into the recycling bin.

People tend to underestimate the most impactful climate actions like voting for leaders who are prepared to fight the climate problem, while overestimating less impactful ones like recycling and replacing lightbulbs.

Getting ten friends to support a climate cause is the single biggest impact you can easily make, and it requires no scientific or technical background. Individuals can use their skills, creativity, and interests to amplify the need to take community action on climate organically and authentically while having fun.

Enormous impacts

- Campaign for political candidates who support climate change initiatives to make sweeping changes to electric, plastic, recycling, and fuel regulations. Campaigns need help that has nothing to do with making donation calls or holding political signs at intersections. Behind the scenes, dozens of people are required to help coordinate schedules, write speeches, and even make food for the other volunteers.
- Submit opinion articles through Citizens Climate Lobby to media platforms and write letters to politicians demanding a nationwide renewable energy standard.
- Email weather reporters encouraging them to discuss climate change when they report on severe storms.
- Join a citizen science project where hikers, skiers, and birdwatchers enter data into a cellphone app to measure snow or tally up birds so scientists can correlate the data to climate change.
- Fundraise for community pedestrian/bike paths and sidewalk projects.
- Find and join climate initiatives aimed at families at the Sierra Club's Climate Parents group.
- Contact local colleges and non-profits that need help with climate change initiatives and research.
- Provide childcare for people who want to rally and campaign for climate initiatives.
- Join a committee that focuses on bringing well-designed mass transit to your community.

Medium impacts

Savings of 2.5+ metric tons of carbon per year

- Choose the train instead of flying short distances. Both take about the same travel time but train travel pollutes significantly less. In Sweden two expressions have become popular on social media: *flygskam* (flight shame) and *tagskryt* (train bragging).
- Sell carbon offsets or credits instead of disposable items like wrapping paper or candy for your organization's next fundraiser.
- Choose electricity providers generating power from the sun and wind.

- Participate in your local "Buy Nothing" group on Facebook to find homes for your castoffs.
- Install water filling stations to cut back on bottled water.
- Listen and share previous generations' tales of thrift and care for the outdoors to foster hope and develop climate care ideas.
- Teach climate change basics at a senior center, scout group, preschool, or library.
- Organize a work carpool with four others.
- Mentor young girls to keep them in school.
- Move your investments away from companies that develop fossil fuels.
- Vote with your wallet in support of sustainable companies that are curbing their emissions and not just "greenwashing."
- Avoid flying for one year and tell others about it in your email signature.
- Organize a curbside composter for the whole condo complex.
- Write an article for your local paper about climate change in your town.

- Tweet your favorite TV chef and urge them to use plant-based meats in their recipes.
- Switch the yearly science fair's theme to center on climate change.

Small impacts

Savings of 0.1-2.4 metric tons of carbon per year

- Save paper
- Recycle
- Replace incandescent bulbs with LEDs
- Shop with reusable grocery bags
- Hang clothing to dry
- Wash clothes in cold water
- Replace the gas car with hybrid or electric
- Telecommute
- Check eBay before buying something new

🌐 **757**

You cannot get through a single day without having an impact on the world around you. What you do makes a difference, and you have to decide what kind of difference you want to make.

— Jane Goodall

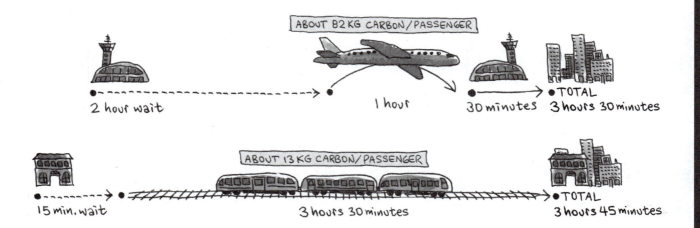

ABOUT 82 KG CARBON/PASSENGER

2 hour wait — 1 hour — 30 minutes — TOTAL 3 hours 30 minutes

ABOUT 13 KG CARBON/PASSENGER

15 min. wait — 3 hours 30 minutes — TOTAL 3 hours 45 minutes

Here's What's True

The science of the climate crisis

What Is Carbon?

An atom is the smallest possible unit of an element. All matter is made up of atoms.

Carbon atoms are the fundamental building blocks of all life on Earth. If there is life outside of Earth, it is also likely that it contains carbon atoms.

Carbon atoms are the fundamental building blocks of all life on Earth.

There are 118 elements on the periodic table. Carbon is one of them. It is the sixth element on the table and represented by a capital C. An element is a building block of the physical universe and cannot be easily broken down further into other components.

Every living thing, plant or animal, is primarily made of atoms of just four elements: carbon, oxygen, hydrogen, and nitrogen.

Eighteen percent of your body is made up of carbon. Trees are made up of about 50 percent carbon. A fish is made up of about 10-15 percent carbon. Diamonds and graphite (the kind present in pencils) are two well-known materials made of 100 percent carbon atoms.

The amount of carbon stored in all living beings on Earth is massive. When a living being dies, its carbon is released into the environment around it. This is how the carbon cycle continues.

What is a gigaton?

When studying the environment of Earth, especially related to the climate crisis, you'll often see a term of measurement called a 'gigaton.' A gigaton is a measure of mass—the amount of matter or substance that makes up an object—in the International System of Units (e.g. metric system).

Units of mass

MEASURE	EQUIVALENT
1,000 grams (g)	1 kilogram (kg)
1,000 kilograms	1 metric ton (t)
1,000,000 metric tons	1 megaton (MT)
1,000 megatons	1 gigaton (GT)

Giga is a prefix meaning 'billion.' So a gigaton is one billion tons. Because our brains are not very efficient at processing numbers so big, it's important to try to maintain perspective on how much mass a gigaton represents.

A gigaton is equivalent to twice the weight of all 7.7 billion humans on Earth. All the cars in the United States, added together, weigh half a gigaton.

⊕ **011**

Not everything that is faced can be changed, but nothing can be changed until it is faced.

— James Baldwin

Carbon is a component of wood, plastic, food, ceramics, steel, and every living thing on Earth.

Carbon in all life forms on Earth

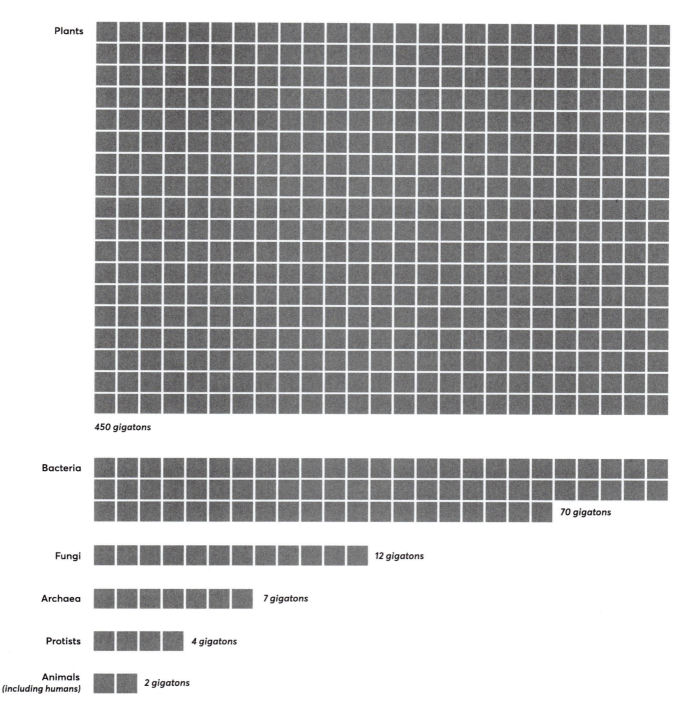

Plants — 450 gigatons

Bacteria — 70 gigatons

Fungi — 12 gigatons

Archaea — 7 gigatons

Protists — 4 gigatons

Animals (including humans) — 2 gigatons

Natural Sources of Carbon Dioxide in the Atmosphere

We can never eliminate CO_2 entirely from our atmosphere—it's actually *supposed* to be there. Natural processes which emit CO_2 include:

Decaying organic matter on land and in the ocean
During decomposition, bacteria and fungi break down urine, feces, and dead organisms into simpler carbon forms such as CO_2.

Volcanic activity
Per a 2019 estimate, 0.28 to 0.36 gigatons of CO_2 enters the air annually when volcanoes erupt or vent, freeing a small portion of the trillions of tons contained beneath the Earth's crust.

Naturally-occurring wildfires
Wildfires emitted an estimated 1.76 gigatons of CO_2 in 2021.

Human respiration
We breathe out CO_2 all day. But those CO_2 emissions don't count as harmful. The reason? That carbon dioxide comes from the photosynthesis of plants, either directly (when we eat grains and produce) or indirectly (when we eat the meat of livestock raised on plants). The same idea also holds true for animal respiration.

From almost a million years, until 1958, natural atmospheric CO_2 levels fluctuated within a range of about 175 to 300 parts per million.

Prior to the Second Industrial Revolution of the mid-1800s—in other words, before the global push for manufacturing and fossil fuels began in earnest—CO_2 levels crept up slowly, taking about 20,000 years to increase 51 percent. By comparison, CO_2 levels have increased 32 percent since 1958, from 316 ppm to 417 ppm.

⊕ **028**

> *We're in a giant car heading towards a brick wall and everyone's arguing over where they're going to sit.*
>
> — David Suzuki

How Much Carbon Are We Talking About?

This image has 10,000 dots on it. If the 10,000 dots represent a breath of air, it turns out that only three of those dots marked to the left are carbon. The rest are oxygen, nitrogen, argon, and small amounts of several other elements.

Actually, it had been three or less for hundreds of thousands of years. It was just over three when regular carbon monitoring started in 1958.

Now it's over four dots. When it gets to five dots, human civilization as we know it will be completely transformed.

That's why this all matters—because the climate is unbelievably sensitive to carbon. One dot out of 10,000 changes everything.

⊕ **336**

43

What Is the Carbon Cycle?

The carbon cycle is the cyclical journey carbon takes from the atmosphere to the Earth and back again.

The total amount of carbon in our system never changes. However, the location and distribution of carbon is constantly changing.

The slow cycle

Carbon moves between rocks, soil, the ocean, and the atmosphere. This process naturally takes between 100 and 200 million years.

Some of the ancient stored carbon becomes rock, while other carbon becomes fossil fuels like coal, oil, and natural gas. We call them "fossil fuels" because dead plants and animals captured underground eventually transform into oil and coal. It takes a very long time.

Volcanoes naturally release stored carbon into the atmosphere by erupting, but so do human activities like burning fossil fuels. Today, humans release 60 times more carbon by burning fossil fuels than all the volcanoes release on Earth each year.

This overabundance of fossil fuel emissions is throwing off the natural balance of carbon in the atmosphere. Human intervention has transformed a slow cycle process into a fast cycle process.

The Fast Cycle

Sometimes carbon exchange happens quickly, as if the Earth is breathing.

Photosynthesis: Carbon dioxide is absorbed by plants and transformed into sugars and oxygen, which is necessary for human life.

Consumption: When animals—including humans—eat plants, the carbon inside the plant is transferred to the animal as energy.

Respiration: Animals breathe in oxygen, which combines with the carbon atoms in our bodies to produce carbon dioxide that is then released back into the atmosphere.

Decomposition: After a plant or animal dies, carbon compounds are broken down; as a result, carbon dioxide is released back into the atmosphere.

On average, the fast carbon cycle takes ten years, with breathing taking a few moments and life and decomposition taking up to a century.

🌐 012

Fast Cycle

10 years

Slow Cycle

200,000,000 years

Balance in the Earth's Carbon Cycle

Carbon dioxide is essential for life on Earth. It absorbs and radiates heat gradually over time, keeping the planet at a livable temperature for humans.

Decomposition of organic matter, breathing (and off-gassing) by plants and animals on land and in the ocean, volcanic eruptions, and wildfires are natural sources of carbon dioxide released into the atmosphere. They are offset by "sinks," which absorb carbon dioxide. The ocean, photosynthesis by land and aquatic plants, and the buildup of soil and peat area are all considered "sinks."

For millions of years, these two forces of the carbon cycle have maintained the atmospheric carbon dioxide level at or below 300 parts per million (ppm)—the three dots per 10,000 discussed above. Levels of carbon dioxide remained stable as the natural sinks removed from the atmosphere an equivalent amount of carbon dioxide created by natural sources.

Ice core samples and modeling show that for the past million years, the Earth has been handling approximately 700 gigatons of carbon dioxide release while staying in natural balance. This natural balance has been destabilized by the addition of carbon dioxide from human activities far faster and at greater levels than there are sinks to remove it.

Imagine trying to add just three more gallons to an already full 75-gallon rain barrel. The barrel is large and a few more gallons seem like a comparatively small addition, but the barrel can't handle it and will overflow. The Earth's carbon cycle, like the rain barrel, is being overwhelmed by the addition of human sources of carbon dioxide.

Without the ability to handle the addition of human source amounts through the opposing sources and sink processes, the small difference that is created in the Earth's natural balance of carbon dioxide is causing the climate disruptions experienced today.

⊕ 029

Carbon-cycle imbalance

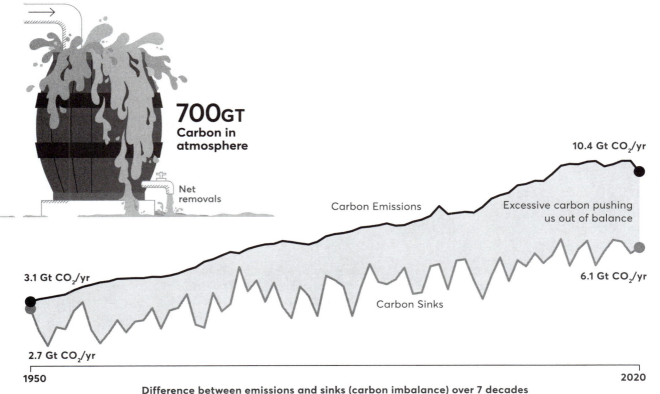

Emissions

700GT
Carbon in atmosphere

Net removals

10.4 Gt CO₂/yr

Carbon Emissions

Excessive carbon pushing us out of balance

3.1 Gt CO₂/yr

6.1 Gt CO₂/yr

Carbon Sinks

2.7 Gt CO₂/yr

1950   2020

Difference between emissions and sinks (carbon imbalance) over 7 decades

M. B. GLASER
Manager
Environmental Affairs Programs

November 12, 1982

CO_2 "Greenhouse" Effect

❝

These models indicate that an increase in global average temperature of $3° \pm 1.5°C$ is most likely.

PAGE 13

One cannot rule out, in view of the inherent uncertainty of the major fluxes, that the biosphere may be a net sink and the oceans may absorb much less of the man-made CO_2.

PAGE 11

The rate of forest clearing has been estimated at 0.5% to 1.5% per year of the existing area. Forests occupy about $50 \times 10^6 km^2$ out of about $150 \times 10^6 km^2$ of continental land, and store about 650 Gt of carbon. One can easily see that if 0.5% of the world's forests are cleared per year, this could contribute about 3.0 Gt/a of carbon to the atmosphere. Even if reforestation were contributing significantly to balancing the CO_2 from deforestation, the total carbon stored in new trees tends to be only a small fraction of the net carbon emitted. It should be noted, however, that the rate of forest clearing and reforestation are not known accurately at this time. If deforestation is indeed contributing to atmospheric CO_2, then another sink for carbon must be found, and the impact of fossil fuel must be considered in the context of such a sink.

PAGE 11

Although all biological systems are likely to be affected, the most severe economic effects could be on agriculture. There is a need to examine methods for alleviating environmental stress on renewable resource production — food, fiber, animal, agriculture, tree crops, etc.

PAGE 21

There is a need to be sure that "lifetime" exposure to elevated CO_2 poses no risks to the health of humans or animals. Health effects associated with changes in the climate sensitive parameters, or stress associated with climate related famine or migration could be significant, and deserve study.

PAGE 21

CO_2 induced warming is predicted to be much greater at the polar regions. There could also be positive feedback mechanisms as deposits of peat, containing large reservoirs of organic carbon, are exposed to oxidation. Similarly, thawing might also release large quantities of carbon currently sequestered as methane hydrates. Quantitative estimates of these possible effects are needed.

PAGE 21

Our best estimate is that doubling of the current concentration could increase average global temperature by about $1.3°$ to $3.1°C$. The increase would not be uniform over the earth's surface with the polar caps likely to see temperature increases on the order of $10°C$ and the equator little, if any, increase.

PAGE 4

Along with a temperature increase, other climatological changes are expected to occur including an uneven global distribution of increased rainfall and increased evaporation. These disturbances in the existing global water distribution balance would have dramatic impact on soil moisture, and in turn, on agriculture.

PAGE 19

In addition to the effects of climate on global agriculture, there are some potentially catastrophic events that must be considered. For example, if the Antarctic ice sheet which is anchored on land should melt, then this could cause a rise in sea level on the order of 5 meters. Such a rise would cause flooding on much of the U.S. East Coast, including the State of Florida and Washington, D.C. The melting rate of polar ice is being studied by a number of glacialogists. Estimates for the melting of the West Anarctica ice sheet range from hundreds of years to a thousand years. Etkins and Epstein observed a 45 mm raise in mean sea level. They account for the rise by assuming that the top 70 m of the oceans has warmed by $0.3°C$ from 1890 to 1940 (as has the atmosphere) causing a 24 mm rise in sea level due to thermal expansion. They attribute the rest of the sea level rise to melting of polar ice. However, melting 51 Tt (10^{12} metric tonnes) of ice would reduce ocean temperature by $0.2°C$, and explain why the global mean surface temperature has not increased as predicted by CO_2 greenhouse theories.

PAGE 19

Atmospheric monitoring programs show the level of carbon dioxide in the atmosphere has increased about 8% over the last twenty-five years and now stands at about 340 ppm. This observed increase is believed to be the continuation of a trend which began in the middle of the last century with the start of the Industrial Revolution. Fossil fuel combustion and the clearing of virgin forests (deforestation) are believed to be the primary anthropogenic contributors although the relative contribution of each is uncertain.

PAGE 4

Jean Senebier's Carbon Discovery

All plants, trees, and algae naturally trap and store carbon dioxide. This is due to photosynthesis, the process in which plants convert sunlight, water, and carbon dioxide into oxygen and plant matter.

Photosynthesis was first discovered in 1782 by Swiss pastor and naturalist Jean Senebier. When plants are exposed to sun, they absorb what Senebier called "fixed air" (carbon dioxide) and emit "good air" (oxygen). Plants do not produce oxygen without both carbon dioxide and sunlight.

At the time, Senebier knew almost nothing about how the process worked, and in fact, it was over a hundred years before scientists discovered rubisco, an enzyme that permits plants to break down carbon dioxide into carbon and oxygen.

Senebier was among the first to suggest plants break down carbon dioxide from the air and store carbon. In perhaps the earliest description of the carbon cycle, Senebier wrote in his *Physiologie Vegetale*:

Dead plants decompose into the ground their debris which form the largest part of the fertiliser from their fermentation. In this way they render to the soil and to the air what they have taken.

Without carbon, plants can't grow. Like humans, they are made of carbon and sequester it in their trunks, their branches, and their roots. And when they die, much of that carbon becomes part of the soil, furthering the cycle.

🌐 **351**

A CARBON PIONEER

In 1856, three years before the work of John Tyndall, American scientist Eunice Foote presented a paper in which she showed that a jar containing carbon dioxide absorbed more heat from the sun than one containing air. She wrote:

An atmosphere of that gas would give to our earth a high temperature; and if as some suppose, at one period of its history the air had mixed with it a larger proportion than at present, an increased temperature from its own action as well as from increased weight must have necessarily resulted.

Plant trees with your web searches.

Find out how at **www.thecarbonalmanac.org/search**.

Carbon Dioxide on Earth over Time

In the past 60 years, the annual rate of increase of carbon dioxide in the atmosphere is approximately 100 times higher than the average rate of increase at the end of the last ice age 11,000 years ago.

Present in the atmosphere since the Earth's creation, levels of carbon dioxide have varied greatly over the planet's 4.54 billion years. These varying levels have caused dramatic fluctuations in the planet's average temperature.

500 million years ago, when atmospheric concentrations of CO_2 were as high as 3,000 to 9,000 ppm, it is estimated that temperatures were more than 14 degrees Celsius higher than 1960-1990 averages.

When the first forms of life emerged 2.5 billion years ago, carbon dioxide began to be consumed through photosynthesis. As life continued to develop, it had a transformative effect on the atmosphere. By 20 million years ago, carbon dioxide levels had dropped to around 300ppm.

Over the last 800,000 years, the Earth has had a regular rhythm of CO_2 levels, with a consistent rise and fall between 150 ppm and 300. In the last 50 years, the rhythm has changed dramatically, and significantly higher CO_2 levels have been recorded over this period.

Natural sources of carbon dioxide are released into the Earth's atmosphere through decomposition of organic matter, breathing (and off-gassing) by ocean and land plants and animals as well as volcanic eruptions and wildfires. Carbon dioxide is absorbed by sinks, processes that bring it into the ocean, photosynthesis by land and aquatic plants, and buildup of soil and peat.

These processes have been keeping the Earth in a natural balance of at or below 300 ppm of atmospheric carbon dioxide throughout the multiple ice ages of at least the past million years.

Viewed on the time scale of Earth's entire geologic history, the dramatic rise in carbon dioxide since the acceleration of carbon-related human activity in the 1750s looks as if it happened overnight. The fossil fuels that have been burned since the start of the Industrial Revolution were created from millions of years of carbon dioxide absorption through plant photosynthesis, yet were sent back into the atmosphere in less than 300 years.

⊕ **030**

Carbon dioxide on Earth over time

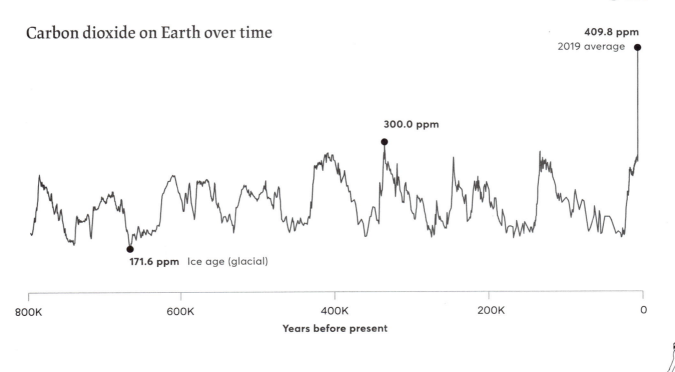

409.8 ppm
2019 average ●

300.0 ppm

171.6 ppm Ice age (glacial)

800K   600K   400K   200K   0

Years before present

Twenty percent of the human body is made of carbon.

Temperature Change on Earth

Accurate thermometer-based temperature records go back to the 1850s. For a measure of temperature on the Earth's surface before that, we use a set of indirect indicators that include the isotopic composition of snow, coral, and stalactites. Layers of snow in the Arctic, for example, can reveal annual weather patterns from long ago.

In addition to this, *dendroclimatology* uses the width of the rings of trees to determine the temperature at a certain time in the past. But the oldest trees on the planet are only a few thousand years old and therefore provide a limited glimpse back in time.

For historic records of the past millions of years, the only reliable tools available are polar ice cores, which can be studied for indications of temperature. The *Milankovitch Cycles* are the cyclical rise and fall

of global temperatures over time. Milankovitch was a Serbian scientist who showed that the cyclical shifts in the Earth's axial tilt, eccentricity, and solar radiation caused these variations.

Geological changes typically take place over very long time frames. However, temperature data from 1880-2020 shows that a substantial change has taken place in the matter of a century.

It is abnormal for a one-degree temperature shift to occur in just one century. Typically such changes take several millennia to take effect.

As Earth's temperature has risen over the last fifty years, Earth has received less light energy from the sun. The Earth's temperature increase isn't caused by the sun. It's due to greenhouse gases trapping more of the sun's heat.

⊕ **366**

Global average temperature change
Compared to the 1961-1990 annual average

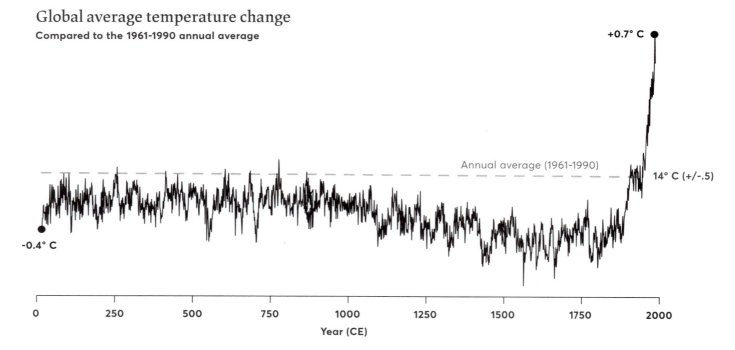

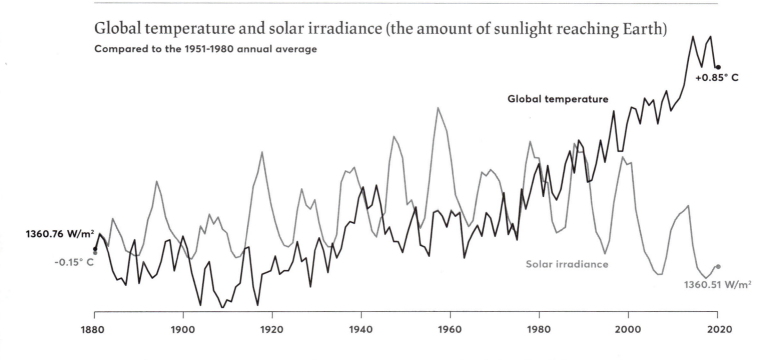

Global temperature and solar irradiance (the amount of sunlight reaching Earth)
Compared to the 1951-1980 annual average

Global temperature

+0.85° C

1360.76 W/m²
-0.15° C

Solar irradiance

1360.51 W/m²

1880 1900 1920 1940 1960 1980 2000 2020

FROM LUSH TO NOT-SO-MUCH

If we scale the Earth's 4.6 billion years to 46 years, then the Industrial Revolution started one minute ago. In that one minute, humans have destroyed more than 50 percent of the world's tropical forests.

CO$_2$ Equivalents

We call the molecules that change our climate "greenhouse gases" (GHG). We often use that term interchangeably with carbon dioxide (CO$_2$) because CO$_2$ is so prevalent.

While CO$_2$ is the most abundant greenhouse gas, other greenhouse gases created by human activity are more powerful or last longer in the atmosphere. Many of these gases exist in trace amounts, but they have an outsized impact on climate change.

To equate the impact of various greenhouse gases with carbon dioxide, their "global warming potential" (GWP) is discussed. For example, over a twenty-year period, methane has more than 80 times the GWP of CO$_2$. That means one ton of methane in the atmosphere is equivalent to 80+ tons of CO$_2$ for the first twenty years after it is emitted.

The other primary human-created greenhouse gases are:
- **Methane:** Produced by cattle and rotting organic material (like food) in landfills. Natural gas consists of mostly methane, so natural gas leaks add methane to the atmosphere.
- **Nitrogen oxides:** Produced mainly by combustion, industrial activities, vehicle exhaust, and fertilizer production, NO$_x$ GWP is 270 times greater than CO$_2$.
- **Fluorinated gases:** Inorganic man-made gases used mostly for refrigeration. They have a GWP more than 1,000 times that of CO$_2$ and also deplete the ozone layer.

⊕ **370**

Note: In this book, CO$_2$ describes the impact of all greenhouse gases, and we use CO$_2$e when specifically describing the impact of other gases.

51

The Relationship Between Population Growth and Emissions

In 1798, the English economist Thomas Malthus posed the question: How many human beings can the Earth support?

The paradox he described—an increase in food production results in population increase greater than the food available—created speculation about human population and its natural limits.

The counterargument to the so-called Malthusian Trap is that more people make more innovations. Greater innovation pushes the practical limits on population further outward. One of these innovations has been using fossil fuels to create energy and fertilizer.

Since Malthus published his book, the global population has increased from fewer than one billion people to more than seven billion, and global living standards have dramatically increased. The amount of carbon burned per person continues to increase as well.

According to Malthus's calculations, good farming practices would lead to grain production doubling in 25 years. But production increases in subsequent 25-year spans would never exceed the increase of the first 25 years. This is called a linear or arithmetic increase.

At the same time, population growth would double in 25 years, and then again double in the next 25 years, and so on. This is called geometric or exponential growth. The rate of population growth would always outpace food production.

The world's population has continued to grow, but the catastrophe Malthus predicted has been avoided again and again. Malthus failed to consider how technology, fueled by fossil fuels, could increase crop yields, health, and productivity.

In 1948 ecologist William Vogt wrote *Road to Survival*, in which he reintroduced Malthus's findings. He said, despite technology, no species can forever exceed the environment's *carrying capacity*. This phrase initially referred to the maximum load a cargo ship could hold. He predicted mankind would surpass the carrying capacity of the planet and its resources, resulting in destruction. Five years later, Eugene P. Odum continued exploring the theme of *carrying capacity* in his *Fundamentals of Ecology.* He established the foundation that led to the concept of planetary boundaries that growing populations needed to be aware of.

However, in the last few decades, innovations in genetics, energy generation, and transportation have led to technological progress that has expanded the carrying capacity of our planet time and again.

The side effects of burning carbon were not in Malthus's equation, however. The increases in production and population over the past two centuries have generated carbon emissions that threaten the quality of life on Earth.

Worldwide, the typical human produces about four tons of carbon dioxide over a lifetime. The range varies wildly: the average person in the United States creates more than 40 times as much carbon as the average person in Bangladesh.

Scientists assert that the maximum carrying load of the Earth is 10 billion humans. According to estimates, this number will be reached by 2050.

Alternatively, some experts insist the maximum estimate should not be based on the capacities of *current* technology. They argue the best way to prepare is for more people to produce more technology, and a rise in population is a positive step.

⊕ **344**

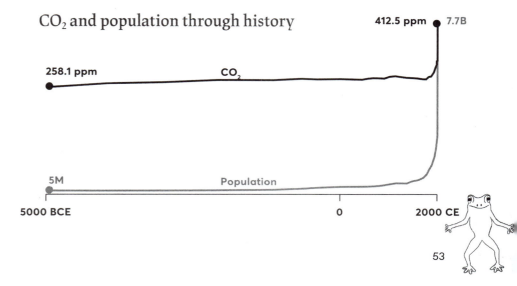

CO$_2$ and population through history

412.5 ppm • 7.7B

258.1 ppm   CO$_2$

5M   Population

5000 BCE   0   2000 CE

53

The History of Systematic Measurement of CO_2

For hundreds of years people have suspected that carbon dioxide impacted the climate, and in the last sixty years scientists have rigorously documented it.

In the 1820s, Joseph Fourier proposed that atmospheric gases might trap heat. A few decades later, John Tyndall conducted experiments demonstrating that carbon dioxide and methane could indeed do so.

Swedish chemist Svante Arrhenius (1859-1927) noticed that burning coal accelerates the greenhouse effect. He was the first to use principles of chemistry to predict the rise of CO_2 in the air and its link to the planet's increasing surface temperature.

English steam engineer and inventor Guy Callendar (1898-1964) further developed Arrhenius' theory. Callendar proved that rising CO_2 concentrations in the atmosphere are linked to rising global temperatures. He was the first to demonstrate that the Earth's land temperature had increased over the previous century. He lived long enough to see his work accepted by climate scientists.

Canadian physicist Gilbert Plass (1920-2004) also predicted the increase in global carbon dioxide levels and its effect on the average temperature. His 1956 predictions are close to the precise measurements made 50 years later. He predicted that a doubling of CO_2 would warm the planet by 3.6°C/6.5°F, that CO_2 levels in 2000 would be 30 percent higher than in 1900, and that the planet would be about 1°C/1.8°F warmer in 2000 than in 1900. Plass said that the spreading envelope of carbon dioxide around the Earth would serve as a greenhouse.

In 1957 US oceanographer Roger Revelle and chemist Hans Suess proved that seawater could not absorb all the additional CO_2 entering the atmosphere. As Revelle put it, "Human beings are now carrying out a large-scale geophysical experiment."

But the first systematic measuring began in 1956 when Charles David Keeling (1928-2005) began his research on CO_2 levels in the atmosphere. He was working to build a device to measure carbon dioxide in the air and discovered significant variations in CO_2, probably due to nearby industry.

He relocated to a dormant volcano in Hawaii, which was more than 4,000 miles downwind from the nearest sources of industrial pollution. At first, the obsessive accuracy of his device was questioned—it was far more precise than the expected changes in the environment he was seeking to measure. But he persisted and painstakingly measured carbon in the atmosphere from the same place and at the same times, year after year for decades.

His method demonstrated that the concentration of CO_2 in the atmosphere was consistently and rapidly increasing. The ongoing project at Mauna Loa, Hawaii, provides the first directly measured, precise proof that CO_2 concentrations are rising.

⊕ 035

CO_2 ppm, 1958–present

Measurements taken at Mauna Loa Observatory

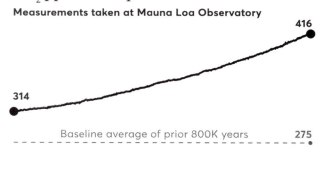

What Is an Ecosystem?

Before 1900, most people viewed each living thing on Earth as a separate entity. As ecological science developed, the concept of interrelated ecosystems became better known.

Ecosystems are geographic areas as large as a rain forest or as small as a tide pool. They can be terrestrial, marine, or freshwater. Within them, a complex relationship exists between living (biotic) populations and their non-living (abiotic) surroundings.

The Earth's entire surface is made up of a web of interconnected ecosystems. Ecosystems can be grouped into categories called *biomes* based on their vegetation, soil, climate, and wildlife. Common biomes include mountains and forests.

Energy flows and matter cycles through an ecosystem. Energy enters the system from the sun and dissipates as heat. Light enters plants through photosynthesis and helps create more matter. Matter is eaten, digested, excreted, decomposed, and eventually eaten again. Tracking these flows and cycles helps us understand how an ecosystem maintains equilibrium and what causes it to become stressed.

Elements of matter like carbon, nitrogen, and phosphorus remain in a continuous cycle through both plants and animals. Plants consume them from the air, water, and soil. Animals consume them by eating other organisms.

In an ecosystem everything is interconnected—from bacteria decomposing parts of the rain forest to lions hunting on the savanna. As a result, every living and non-living thing impacts the balance of their ecosystem, as well as systems beyond their own. If one thing changes or disappears, everything else in the chain will feel the impact.

When rabbits were introduced to Australia, the life of every other creature was impacted. When large predators become extinct, the animals they used to hunt become more numerous, impacting the growth of plants.

And the same is true for energy. As temperatures rise, living communities can shift and become out of sync with each other.

🌐 **352**

Planetary Boundaries: Limits of the Natural World

In 2009, Johan Rockstrom and a team of researchers listed nine processes that regulate the stability of our climate. Each one is unpredictable—instead of a strictly linear response, when pressed in response to inputs they may produce a "cliff." In addition, some of these factors interact with the others, further worsening the situation.

1. **Climate change:** As the planet warms, it accelerates further warming. Forests and oceans that absorb carbon are also harmed by rising temperatures, meaning that instead of absorbing more carbon, they become less effective. Rainfall leads to flooding, which leads to topsoil erosion, which can lead to a rise in temperatures.

2. **Biodiversity loss and species extinction:** As species become extinct, other species multiply, harming the vegetation they depend on, which can lead to more species becoming extinct.

3. **Stratospheric ozone depletion:** As the ozone layer decreases, it causes human impacts, but it also harms marine life and accelerates climate change.

4. **Ocean acidification:** As the oceans absorb CO_2, carbonic acid is created. This changes ocean chemistry and increases its acidity. This impacts the biosystem and decreases the ability of the ocean to absorb more carbon.

5. **Biogeochemical flows of phosphorus and nitrogen:** More warming pushes farmers to use more fertilizer, which runs off into rivers and causes algae blooms. In turn, this degradation can lead to less carbon sequestration (storage) and a rise in temperatures.

6. **Deforestation and other land-system change:** As farms undergo stress, it's more likely that natural lands including forests and marshes will be converted to agricultural land. This decreases biodiversity, changes water flows, and dramatically impacts carbon sequestration.

7. **Decreasing availability of fresh water:** As water supplies decrease, humans will take more dramatic measures to obtain more water, causing stress on the system and impacting health.

8. **Introduction of novel entities:** Synthetic organic pollutants, heavy metal compounds, and radioactive materials are more likely to be released, which poison the atmosphere and cause problems with human and animal health.

9. **Atmospheric aerosol loading:** Increases in water vapor change the mechanics of atmospheric aerosol loading. This can change the impacts of solar radiation in unpredictable ways.

⊕ **339**

Understanding the Greenhouse Effect

It's said that the Roman emperor Tiberius insisted on having a cucumber every day for breakfast. In order to provide this for him year-round, his staff created the original greenhouse.

A greenhouse is a glass building that allows sunshine in while also keeping heat from escaping. It permits gardeners to grow plants even when it's cold outside. The glass permits light to enter, and the infrared radiation in the sunlight heats up the plants, the air, and the pots.

A greenhouse gas does almost the same thing to our planet as the glass does to the greenhouse. Carbon and other molecules far above the Earth's surface permit sunshine to reach us, but act as a blanket, keeping the heat in the atmosphere from escaping into space.

That's the metaphor. The actual physics are a bit different. A greenhouse uses glass to physically keep air from moving. Throughout our atmosphere, tiny amounts of carbon dioxide and other greenhouse gases keep some of the infrared heat given off by the Earth from escaping into outer space.

If the air was simply 100 percent oxygen and nitrogen (instead of only 99 percent), there would be little retained heat in our atmosphere. Add water vapor, and some of the infrared heat remains. Add carbon dioxide and other greenhouse gases, and we reach a tipping point in the heat that is retained.

Even though the metaphor is inexact, it's a useful way to think about the dynamic. An invisible substance provides a barrier of sorts that keeps heat in.

Without some sort of greenhouse effect provided by the atmosphere, the Earth would be as cold as space and no life would be possible. But when we add more greenhouse gases (like carbon dioxide, methane, and others) to the air, the balance changes. The rapidly changing imbalance in our atmosphere is the result of a change in the percentage of greenhouse gases that are in the air. And so, our planet, on average, is getting hotter.

🌐 **355**

Without us, Earth will abide and endure; without her, however, we could not even be.

— Alan Weisman

Before humans began burning fossil fuels, heat easily escaped the atmosphere.

As greenhouse gases accumulate, more heat is reflected back into the atmosphere.

Why Are the Greenhouse Emissions Numbers So Confusing?

Do the numbers add up?

The editors of this Almanac have reviewed more than 5,000 data sources, and if your experience is like ours, the numbers can get confusing. We regularly see that one human activity or another is responsible for "more than 8 percent of all carbon emissions." Looking at the data, you might expect to see wildly divergent predictions of our future, but it's surprising to see different statements about the present.

Looking beyond simple charts to interpret greenhouse emissions data is the first step. It's also important to consider the sources behind the data, to read the footnotes, and to understand that certain narratives can be supported by unintentional or deliberate data manipulation. There's inadvertent double-counting, challenges with co-dependent variables and confusion about labels and definitions.

Obtaining raw data from a trusted source is ideal. In the absence of raw data, some authorities are more impartial and transparent than others. In the climate field, researchers often allow their data to be downloaded so people can analyze it themselves and draw their own conclusions. As a result, sometimes the same data can advance different narratives.

Same data, different pictures

One common problem we found is the challenge of determining where in the production chain the data is pointing. A hamburger, for example, needs meat, and meat needs a truck, and a truck picks up a cow, and a cow needs a field, and a field needs fertilizer... so does the fertilizer's production count as carbon related to food?

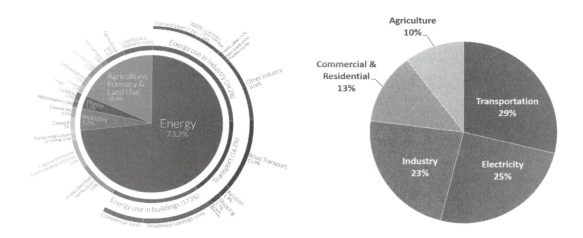

Two popular online charts. The one on the left is hopelessly complex, assuring the reader that the sources of carbon dioxide are too intertwined to understand. The one on the right oversimplifies, incorrectly sorting things into buckets without regard for overlap. There's no easy, obvious, and correct way to represent this system in one chart.

For example, in the two charts, greenhouse gas emissions are attributed using different descriptors and varying levels of detail.

The first chart attributes the greenhouse gas emissions by sector and drills it down further to the final *energy consumer*. Breaking down the data in such a way can provide a clearer picture of emissions by industry.

The second chart uses more generic labeling that leaves it unclear whether it's attributed to the *consumer* of the energy or the *producer* of it.

Follow the chain long enough, and you'll end up with the top oil companies—which produce huge amounts of the carbon that is released, but don't actually use most of it. They sell it.

Neither of these charts is incorrect from a data perspective, but they each appear to tell a different story that may lead readers to reach different conclusions.

Complex calculations

The chart below illustrates a number of direct and indirect activities that impact the corporate reporting of greenhouse gas emissions. Each one plays a role in the process of creating and delivering a product or service.

With so many steps in the process, getting an accurate measure of true emissions can be pretty challenging. Where does one activity start and another end? How do emissions get recorded and by whom?

If a power plant burns recycled plastics for energy, is it green? It depends on whether you count the plastic's carbon impact as over and done with before the plant gets to it.

A corporation's staff doesn't have to work very hard to make the company sound carbon neutral. They simply have to categorize as many carbon activities as indirect as they can.

⊕ **023**

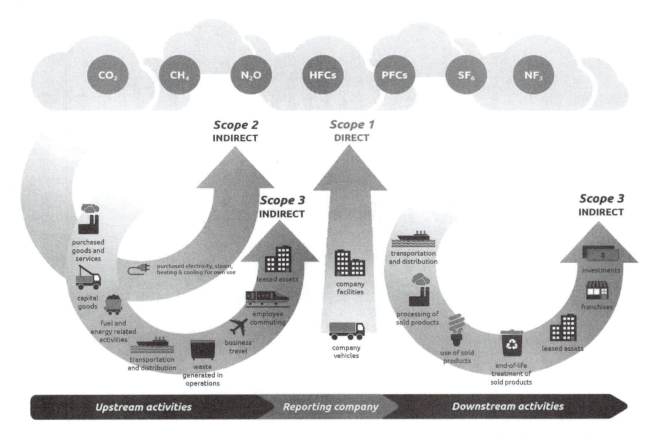

More graph confusion: This chart, found many places online, explains how companies can easily obfuscate their impact on the environment. The creator of the chart wanted to show how easy it was to confuse people, and succeeded.

Every 4,434 metric tons of CO_2 pumped into the atmosphere results in one premature death.

Our Choices Can Have a Lethal Impact

Lifestyle choices are a significant contributor to carbon emissions in the atmosphere. Rates of consumption vary from country to country.

Researchers at *Nature* calculated that for every 4,434 metric tons of CO_2 pumped into the atmosphere (beyond the 2020 rate of emissions) one more premature death from increased temperature will result.

To put this into context, it takes only 3.5 average US citizens to create enough emissions within a lifetime to materially shorten the life of another person. By comparison, it would take 25 Brazilians or 146 Nigerians to do the same.

⊕ **341**

How many people does it take?

CO₂ PRODUCED

3.5
Americans

4,434
metric
tons

25
Brazilians

4,434
metric
tons

146
Nigerians

4,434
metric
tons

Carbon Feedback Cycles

The climate is responding to three forces.

The first force is how much humans emit. This is influenced by factors such as population growth rates, gasoline versus electric cars, shifting power generation away from fossil fuels, political will, etc. Cultural and technological trends are very difficult to predict far in advance, but trends are shifting.

The second force is how sensitive the environment is to our actions. How will the atmosphere and the ocean respond to increases in carbon dioxide and other greenhouse gases? We can make smart assumptions about linear or exponential changes based on historic data and laboratory experiments.

And the third force— the most complex of the three —is how the systems of our environment will interact with each other when confronted with these changes.

Called *carbon-cycle feedbacks*, this force holds a multitude of complex processes that exchange carbon between the atmosphere, land, oceans, plants, and animals. It is what the world does with the greenhouse gases that are introduced and the change in climate that results from the second force.

Consider a child's swing set. It's not difficult to swing back and forth a little. But as you gain momentum, it's more and more difficult to extend the length of travel. It's essentially impossible to swing all the way to the top of the swing set. The closer we get to the end, the harder it is to go farther.

Some systems are *positive feedback* in nature, while others create *negative* feedback. A negative feedback system works to maintain stability. The effort put into moving it out of balance is met with resistance, pushing it back to where it was. A positive feedback system can spiral out of control, with each disruption causing the wobble to get worse, creating more wobbling.

For millennia, when carbon dioxide increased, the ocean absorbed it. When it decreased, the ocean released it. The ocean created a negative feedback system to keep carbon dioxide levels stable.

Currently, about 30 percent of CO_2 emissions are being taken up by the oceans. One force absorbing carbon dioxide is the Atlantic Meridional Overturning Circulation (AMOC), a system of ocean currents. The AMOC is driven by ocean temperatures and salinity, both of which will change in a shifting climate. In the future, the AMOC may not absorb as much CO_2 as it does today.

As a result, fewer emissions would be taken up by the oceans, and the planet would warm faster. In turn, the AMOC could become even less effective. It would change from a stable, negative feedback system that absorbs emissions to one that continually becomes less effective at that task.

Similar carbon-cycle feedback systems occur elsewhere. Long ago, methane was trapped in Arctic permafrost. As the Arctic warms and permafrost thaws, methane gas will be released. The additional methane in the atmosphere will further contribute to temperatures rising at the poles, leading to a cycle of increased permafrost thawing.

Positive feedback loops can become a multiplying force that accelerates the challenges of a changing climate. 🌐 **368**

Climate Change
Temperature, T

Respiration

CO₂ Solubility

Radiative Forcing

Land Carbon
Total vegetation and soils, C_t

Ocean Carbon
Ocean mixed layer, C_m
(Total Ocean), C_M

Fertilization Effect

Carbon Uptake

Partial Pressure

Carbon Uptake

Atmosphere
Carbon, C_a

What Do I Get for 1kg CO$_2$?

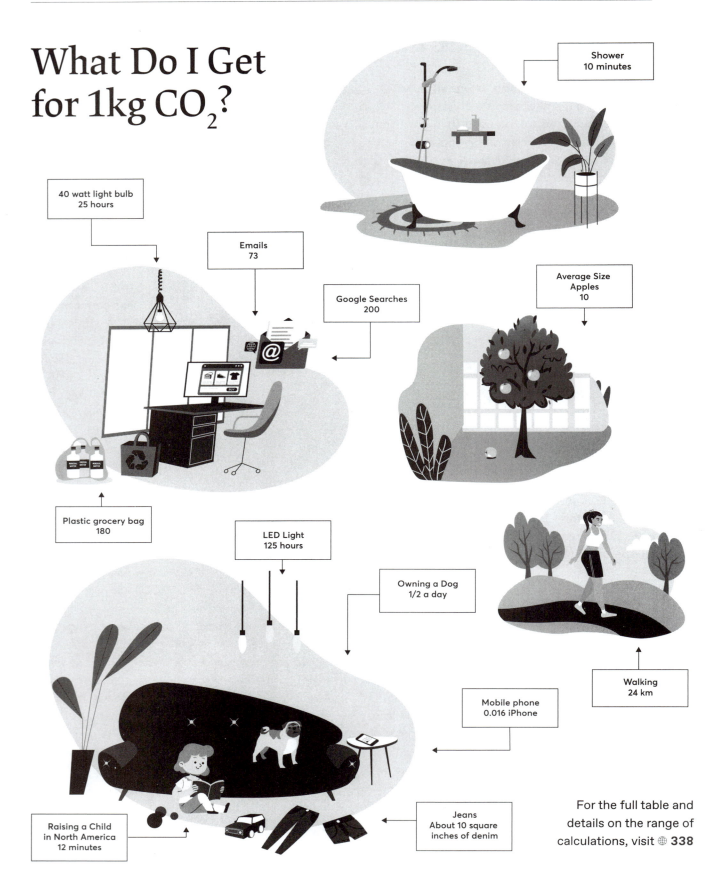

Shower
10 minutes

40 watt light bulb
25 hours

Emails
73

Google Searches
200

Average Size
Apples
10

Plastic grocery bag
180

LED Light
125 hours

Owning a Dog
1/2 a day

Walking
24 km

Mobile phone
0.016 iPhone

Raising a Child
in North America
12 minutes

Jeans
About 10 square
inches of denim

For the full table and
details on the range of
calculations, visit ⊕ 338

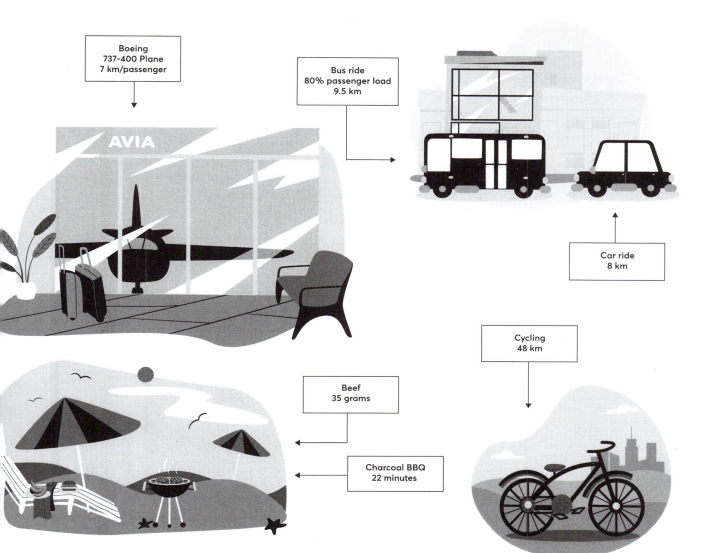

Boeing
737-400 Plane
7 km/passenger

Bus ride
80% passenger load
9.5 km

Car ride
8 km

Cycling
48 km

Beef
35 grams

Charcoal BBQ
22 minutes

Broccoli
2 kg

Average Size
Avocado
2

Average Size
Eggs
5

Cheese
100 g

CO$_2$ Emissions by Country

Cumulative carbon emissions by country (1960-2018)

Cumulative emissions over time reflect the historical responsibility for and impact on climate change by a given country. Although a country may have lowered its emissions over time, its historical emissions are one important factor in who we should hold accountable to solving the global problem today.

	COUNTRY	GIGATONS			COUNTRY	GIGATONS
				25	Argentina	6.86
	World	**1280.93**		26	Belgium	6.57
1	United States	279.25		27	Thailand	6.54
2	China	204.69		28	Venezuela, RB	6.46
3	Russian Federation	131.28		29	Egypt, Arab Rep.	5.64
4	Japan	56.19		30	Kazakhstan	5.5
5	India	45.68		31	Malaysia	5.05
6	United Kingdom	32.4		32	Korea, Dem. People's Rep.	4.82
7	Canada	25.51		33	Pakistan	4.35
8	Germany	23.7		34	United Arab Emirates	4.33
9	France	22.65		35	Iraq	4.07
10	Italy	20.92		36	Algeria	3.93
11	Poland	19.65		37	Hungary	3.73
12	Mexico	17.13		38	Greece	3.6
13	Korea, Rep.	16.62		39	Sweden	3.54
14	South Africa	16.24		40	Bulgaria	3.44
15	Iran, Islamic Rep.	15.77		41	Czech Republic	3.43
16	Australia	15.31		42	Austria	3.39
17	Brazil	13.72		43	Vietnam	3.33
18	Saudi Arabia	12.65		44	Uzbekistan	3.3
19	Spain	12.37		45	Philippines	3.1
20	Indonesia	11.98		46	Denmark	3.06
21	Ukraine	9.61		47	Colombia	2.9
22	Turkey	9.22		48	Finland	2.85
23	Netherlands	8.77		49	Kuwait	2.44
24	Romania	7.21		50	Switzerland	2.36

Carbon emissions by country (2018)

Each country's emissions changes over time. In order to reach net-zero emissions globally, every country's yearly emissions will need to drop from where they are today to zero or near-zero. Current emissions are a good indicator of how far a country has to go.

	COUNTRY	GIGATONS
1	China	10.31
2	United States	4.98
3	India	2.43
4	Russian Federation	1.61
5	Japan	1.11
6	Germany	0.71
7	Korea, Rep.	0.63
8	Iran, Islamic Rep.	0.63
9	Indonesia	0.58
10	Canada	0.57
11	Saudi Arabia	0.51
12	Mexico	0.47
13	South Africa	0.43
14	Brazil	0.43
15	Turkey	0.41
16	Australia	0.39
17	United Kingdom	0.36
18	Italy	0.32
19	France	0.31
20	Poland	0.31
21	Thailand	0.26
22	Vietnam	0.26
23	Spain	0.26
24	Egypt, Arab Rep.	0.25
25	Malaysia	0.24

Carbon emissions per capita by country (2018)

Per capita measures of emissions are a reflection of the intensity of resource use in a given country. On a global scale, total emissions are what led to changing temperatures on Earth. On a national scale, this measure shows on average how much each person in that country emits per year.

	COUNTRY	METRIC TONS
1	Qatar	32.42
2	Kuwait	21.62
3	United Arab Emirates	20.80
4	Bahrain	19.59
5	Brunei Darussalam	16.64
6	Palau	16.19
7	Canada	15.50
8	Australia	15.48
9	Luxembourg	15.33
10	Saudi Arabia	15.27
11	United States	15.24
12	Oman	15.19
13	Trinidad and Tobago	12.78
14	Turkmenistan	12.26
15	Korea, Rep.	12.22
16	Estonia	12.10
17	Kazakhstan	12.06
18	Russian Federation	11.13
19	Czech Republic	9.64
20	Libya	8.83
21	Netherlands	8.77
22	Japan	8.74
23	Germany	8.56
24	Singapore	8.40
25	Poland	8.24

🌐 **018**

Global Greenhouse Gas Emissions by Sector

In 2016, humans were responsible for 49.4 gigatons of CO_2 equivalent emissions

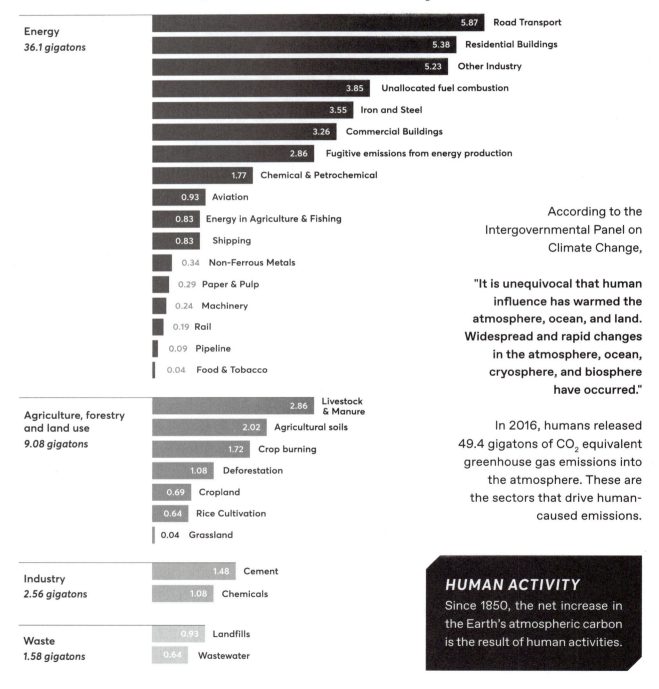

Energy
36.1 gigatons

5.87	Road Transport
5.38	Residential Buildings
5.23	Other Industry
3.85	Unallocated fuel combustion
3.55	Iron and Steel
3.26	Commercial Buildings
2.86	Fugitive emissions from energy production
1.77	Chemical & Petrochemical
0.93	Aviation
0.83	Energy in Agriculture & Fishing
0.83	Shipping
0.34	Non-Ferrous Metals
0.29	Paper & Pulp
0.24	Machinery
0.19	Rail
0.09	Pipeline
0.04	Food & Tobacco

Agriculture, forestry and land use
9.08 gigatons

2.86	Livestock & Manure
2.02	Agricultural soils
1.72	Crop burning
1.08	Deforestation
0.69	Cropland
0.64	Rice Cultivation
0.04	Grassland

Industry
2.56 gigatons

1.48	Cement
1.08	Chemicals

Waste
1.58 gigatons

0.93	Landfills
0.64	Wastewater

According to the Intergovernmental Panel on Climate Change,

"It is unequivocal that human influence has warmed the atmosphere, ocean, and land. Widespread and rapid changes in the atmosphere, ocean, cryosphere, and biosphere have occurred."

In 2016, humans released 49.4 gigatons of CO_2 equivalent greenhouse gas emissions into the atmosphere. These are the sectors that drive human-caused emissions.

HUMAN ACTIVITY
Since 1850, the net increase in the Earth's atmospheric carbon is the result of human activities.

Here's how that breaks down:

*All percentages are measures of the entire amount of emissions released. Source: Our World in Data

Energy

73.2%*

36.1 gigatons

ENERGY USE IN INDUSTRY: 24.2%

- **Iron & Steel (7.2%):** Energy-related emissions from the manufacturing of iron and steel.
- **Chemical & petrochemical (3.6%):** Energy-related emissions from the manufacturing of fertilizers, pharmaceuticals, refrigerants, oil and gas extraction, etc.
- **Food & tobacco (1%):** Energy-related emissions from the manufacturing of tobacco products and food processing (e.g., the conversion of raw agricultural products into their final products, such as the conversion of wheat into bread).
- **Non-ferrous metals (0.7%):** Non-ferrous metals are metals which contain very little iron: these include aluminum, copper, lead, nickel, tin, titanium, zinc, and alloys such as brass. The manufacturing of these metals requires energy which results in emissions.
- **Paper & pulp (0.6%):** Energy-related emissions from the conversion of wood into paper and pulp.
- **Machinery (0.5%):** Energy-related emissions from the production of machinery.
- **Other industry (10.6%):** Energy-related emissions from manufacturing in other industries including mining and quarrying, construction, textiles, wood products, and transport equipment (such as car manufacturing).

TRANSPORT: 16.2%

This includes a small amount of electricity (indirect emissions) as well as all direct emissions from burning fossil fuels to power transport activities (these figures do not include emissions from the manufacturing of motor vehicles or other transport equipment, which falls under "Energy use in Industry").

- **Road transport (11.9%):** Emissions from the burning of petrol and diesel from all forms of road transport, which includes cars, trucks, lorries, motorcycles, and buses. Sixty percent of road transport emissions come from passenger travel (cars, motorcycles, and buses) and the remaining forty percent from road freight (lorries and trucks). This means that, if we could electrify the whole road transport sector, and transition to a fully decarbonized electricity mix, we could feasibly reduce global emissions by 11.9%.
- **Aviation (1.9%):** Emissions from passenger travel and freight and domestic and international aviation. 81 percent of aviation emissions come from passenger travel with the remaining 19 percent from freight. 60 percent of passenger aviation emissions come from international travel and 40 percent from domestic.
- **Shipping (1.7%):** Emissions from the burning of petrol or diesel on boats. This includes both passenger and freight maritime trips.
- **Rail (0.4%):** Emissions from passenger and freight rail travel.
- **Pipeline (0.3%):** Fuels and commodities (e.g., oil, gas, water, or steam) often need to be transported (either within or between countries) via pipelines. This requires energy inputs, which results in emissions (poorly constructed pipelines can also leak, but this falls under the category "Fugitive emissions from energy production").

ENERGY USE IN BUILDINGS: 17.5%

- **Residential buildings (10.9%):** Energy-related emissions from the generation of electricity for things like lighting, appliances, cooking, heating and cooling in homes.
- **Commercial buildings (6.6%):** Energy-related emissions from the generation of electricity for things like lighting, appliances, heating, and cooling in offices, restaurants, shops, etc.

UNALLOCATED FUEL COMBUSTION: 7.8%

Energy-related emissions from the production of energy from other fuels, including electricity and heat from biomass, on-site heat sources, combined heat and power (CHP), nuclear industry, and pumped hydroelectric storage.

FUGITIVE EMISSIONS: 5.8%

- **Fugitive emissions from oil and gas (3.9%):** Fugitive emissions that are the often-accidental leakage of methane to the atmosphere during oil and gas extraction and transportation and from damaged or poorly maintained pipes. This also includes flaring (the intentional burning of gas to prevent its accidental release during extraction).
- **Fugitive emissions from coal (1.9%):** Fugitive emissions that are the accidental leakage of methane during coal mining.

AGRICULTURE AND FISHING: (1.7%)

Energy-related emissions from the use of machinery in agriculture and fishing, such as fuel for farm machinery and fishing vessels.

Agriculture, Forestry, Land Use

18.4%
9.08 gigatons

Agriculture, forestry, and land use directly account for 18.4 percent of all greenhouse gas emissions. The food system as a whole—including refrigeration, food processing, packaging, and transport—accounts for around one-quarter of greenhouse gas emissions.

- **Grassland (0.1%):** When grassland becomes degraded, these soils can lose carbon, converting to carbon dioxide in the process. Conversely, when grassland is restored (for example, from cropland), carbon can be sequestered. Emissions here therefore refer to the net balance of these carbon losses and gains from grassland biomass and soils.
- **Cropland (1.4%):** Depending on the management practices used on croplands, carbon can be lost or sequestered into soils and biomass. This affects the net balance of carbon dioxide emissions (this does not include grazing lands for livestock).
- **Deforestation (2.2%):** Net emissions of carbon dioxide from changes in forestry cover (refor- estation is counted as "negative emissions" and deforestation as "positive emissions"). Emissions are based on lost carbon stores from forests and changes in carbon stores in forest soils.
- **Crop burning (3.5%):** Farmers often burn crop residues—leftover vegetation from crops such as rice, wheat, sugar cane, and other crops—after harvest to prepare land for the resowing of crops. Their burning releases carbon dioxide, nitrous oxide, and methane.
- **Rice cultivation (1.3%):** Flooded paddy fields produce methane through a process called "anaerobic digestion." Organic matter in the soil is converted to methane due to the low-oxygen environment of water-logged rice fields.
- **Agricultural soils (4.1%):** Nitrous oxide—a green- house gas—is produced when synthetic nitrogen fertilizers are applied to soils. This includes emissions from agricultural soils for all agricul- tural products, including food for direct human consumption, animal feed, biofuels, and other non-food crops (such as tobacco and cotton).
- **Livestock & manure (5.8%):** Animals (mainly ruminants, such as cattle and sheep) produce greenhouse gases through a process called "enteric fermentation"—when microbes in their digestive systems break down food, they produce methane as a byproduct.

Nitrous oxide and methane can be produced from the decomposition of animal manures under low-oxygen conditions. This often occurs when large numbers of animals are managed in a confined area (such as dairy farms, beef feedlots, and swine and poultry farms), where manure is typically stored in large piles or disposed of in lagoons and other types of manure management systems (emissions here include direct emissions from livestock only and do not consider impacts of land use change for pasture or animal feed).

Direct Industrial Processes

5.2%
2.56 gigatons

- **Cement (3%):** Carbon dioxide is produced as a byproduct of a chemical conversion process used in the production of clinker, a component of cement. (Cement production also produces emissions from energy inputs that fall under "Energy Use in Industry".)
- **Chemicals & petrochemicals (2.2%):** Greenhouse gases can be produced as a byproduct from chemical processes. For example, CO_2 can be emitted during the production of ammonia, which is used as a refrigerant, for purifying water supplies, in cleaning products, and in the production of many materials, including plastic, fertilizers, pesticides, and textiles. Chemical and petrochemical manufacturing also produces emissions from energy inputs that fall under "Energy Use in Industry."

Waste

3.2%
1.58 gigatons

- **Wastewater (1.3%):** Organic matter and residues from animals, plants, humans, and their waste products can collect in wastewater systems. When this organic matter decomposes it produces methane and nitrous oxide.
- **Landfills (1.9%):** Landfills are often low-oxygen environments. In these environments, organic matter is converted to methane when it decomposes.

🌐 013

Search the web, plant a tree

The Carbon Almanac is teaming up with Ecosia to make your online searching more powerful. Visit **www.thecarbonalmanac.org/search** to install a simple extension that plants a tree every time you do some web searches. It's free. Just as fast and even easier than Google, but it makes a difference, every day.

143,000,000 trees planted as of 2021.

Where Does All the Carbon Go?

For millions of years, the level of carbon dioxide in the atmosphere has fluctuated in a cyclic rhythm. When the level of carbon dioxide rises, plants and oceans absorb more of it. When the level decreases, the opposite occurs. This rhythm changed two hundred years ago with the beginning of the Industrial Age.

Humans now produce about 34 billion tons of carbon each year, a number that has risen exponentially in recent decades. Over a third of that will go into the atmosphere. The current concentration is 412 parts per million (ppm), up from 300 a few generations ago. Over the past 60 years, carbon dioxide has been increasing 100 times faster than previously recorded natural increases. Prehistoric increases and decreases occurred regularly due to natural factors. However, man-made carbon dioxide production is orders of magnitude larger.

Much of the carbon humans emit is absorbed by plants and oceans. But not all of it.

Roughly 25 percent of the carbon will be assimilated by oceans, typically via the atmosphere. Oceans have traditionally been a significant carbon sink, but evidence indicates that they may be reaching a limit.

Plants and soils currently absorb an additional 30 percent of the carbon dioxide that's produced. Changes in the amount of carbon dioxide in the air will change the fertility and growth rates of plants, and some will thrive and grow faster. Researchers estimate that grasslands could absorb larger quantities than initially assumed. But there is no credible evidence that the Earth's natural systems will be able to incorporate all the increases in carbon dioxide in the near future.

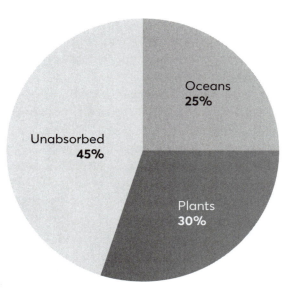

That leaves 45 percent of the carbon produced unaccounted for—a gap that is leading to the shift in the makeup of the atmosphere and the change in the climate.

Work is under way to increase the Earth's ability to capture and store greenhouse gases. Engineers are developing technologies to permit plants, microorganisms, and oceans to store more carbon. In addition, work is being done on acquiring and storing carbon directly from the atmosphere. Carbon sequestration is a new process that promises to capture CO_2 emissions and transfer them underground. Carbon can then be stored, mimicking the slow carbon cycle. However, to date these technologies have been expensive and largely ineffective at adequately addressing the carbon gap.

⊕ **365**

" It is 30 years since we started this work. Activities that devastate the environment and societies continue unabated. Today we are faced with a challenge that calls for a shift in our thinking, so that humanity stops threatening its life-support system.

There is also a need to galvanize civil society and grassroots movements to catalyze change. I call upon governments to recognize the role of these social movements in building a critical mass of responsible citizens, who help maintain checks and balances in society. On their part, civil society should embrace not only their rights but also their responsibilities. . . .

Further, industry and global institutions must appreciate that ensuring economic justice, equity and ecological integrity are of greater value than profits at any cost. The extreme global inequities and prevailing consumption patterns continue at the expense of the environment and peaceful co-existence. The choice is ours. . . .

As I conclude I reflect on my childhood experience when I would visit a stream next to our home to fetch water for my mother. I would drink water straight from the stream. . . . I saw thousands of tadpoles: black, energetic and wriggling through the clear water against the background of the brown earth. This is the world I inherited from my parents.

Today, over 50 years later, the stream has dried up, women walk long distances for water, which is not always clean, and children will never know what they have lost. The challenge is to restore the home of the tadpoles and give back to our children a world of beauty and wonder.

— Wangari Maathai

Energy Production and Carbon

Energy production accounts for 73.2 percent of global CO_2 emissions. In 2019, humans consumed over 160,000 terawatt hours (TWh) of energy. This is eight times more power than the entire world used in 1950, even though the population is only three times bigger. Emissions from energy have grown by more than 700 percent over the same period.

Producing that much energy created more than 15 gigatons of CO_2 emissions. As of 2019, low- or no-carbon sources supply only 15.7 percent of our energy. The rest comes from fossil fuels. Coal, oil, and natural gas produce the most emissions for every unit of energy they create. Coal produces more GHGs per unit of output than any other source. Increased coal use in the last 10 years has outpaced the growth of alternative energy sources. Net-zero emissions by 2050 will require the near or complete elimination of fossil fuels from the energy sector by 2050.

🌐 **020**

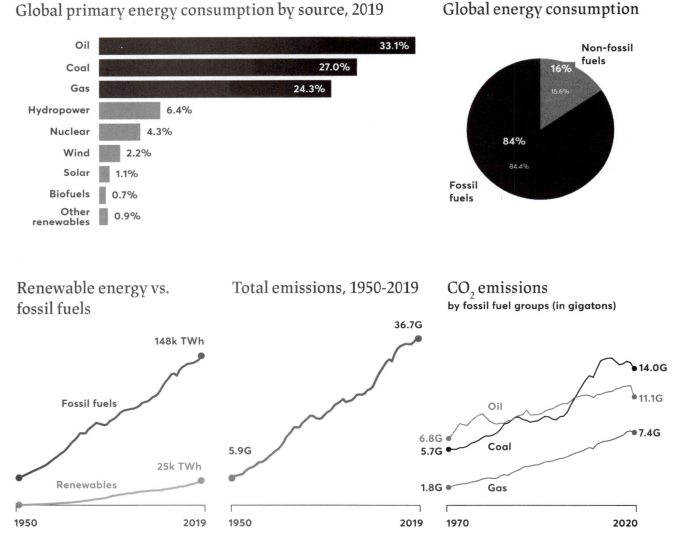

Global primary energy consumption by source, 2019

- Oil — 33.1%
- Coal — 27.0%
- Gas — 24.3%
- Hydropower — 6.4%
- Nuclear — 4.3%
- Wind — 2.2%
- Solar — 1.1%
- Biofuels — 0.7%
- Other renewables — 0.9%

Global energy consumption

- Non-fossil fuels 16% (15.6%)
- Fossil fuels 84% (84.4%)

Renewable energy vs. fossil fuels

- Fossil fuels 148k TWh
- Renewables 25k TWh
- 1950 — 2019

Total emissions, 1950-2019

- 36.7G
- 5.9G
- 1950 — 2019

CO_2 emissions
by fossil fuel groups (in gigatons)

- Oil 6.8G → 14.0G
- Coal 5.7G → 11.1G
- Gas 1.8G → 7.4G
- 1970 — 2020

The Energy Cost of Plug Loads

A century ago, we began installing electrical outlets into buildings. The devices we plug into these outlets can contribute upwards of 50 percent of total energy consumption in a single building. These devices and the energy they use are known as plug loads. When aggregated, these plug loads are responsible for a significant amount of electrical consumption and thus carbon emissions.

As more devices become wired to the Internet, plug loads continue to grow. While efficiencies of some devices are increasing, the need to be on or in standby mode more often increases the energy consumption of each device.

Standby power consumption is generally defined as electricity used by appliances and equipment while they are switched off or not performing their primary function. This is often referred to as vampire or phantom power.

Standby power consumption has been decreasing over time. For example, at the turn of the century, videocassette recorders (VCRs) in the United States used more electricity while in standby mode (i.e., flashing the dreaded 12:00!) than while actively recording or playing. In New Zealand, many microwave ovens consumed more electricity in standby mode powering the clock and keypad than in cooking food.

Seeing this as a design issue, many manufacturers responded. Over the past twenty years, considerable progress has been made in reducing standby power consumption in specific products. For example, the standby power consumption of mobile phone chargers has been reduced from more than two watts in 2000 to below 0.3 watts today.

This single reduction is critical from a carbon perspective given that in 2000 there were only 740 million cell phone subscriptions worldwide but in 2020 there were 8.3 billion.

⊕ **362**

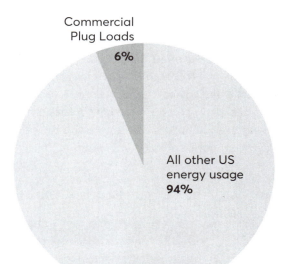

Commercial Plug Loads **6%**

All other US energy usage **94%**

SHARING SOLAR POWER

In Bangladesh, 6 million homes have solar systems, but more than 50 million homes have no reliable access to electricity. Innovative "solar sharing" models allow those who produce solar power to share it with their neighbors and provide access to clean, as-needed electricity for those who otherwise couldn't afford it.

The greenest building is the one that already exists. — Carl Elefante

Construction's Carbon Debt

The total square footage of buildings worldwide is expected to double by 2060 to accommodate population growth, mostly in developing countries. Essentially, the construction sector will add the equivalent of New York City every month for the next 40 years.

The UN goal to reach a zero-carbon building stock by 2050 will require innovation in all industries that contribute to the direct and indirect emissions of the construction sector. However, the Buildings Climate Tracker shows that progress toward this goal actually *declined* between 2016 and 2019.

The building and construction sector generates 38 percent of annual global CO_2 emissions. 74 percent is related to energy use, and 26 percent comes from the materials used to build the structure and the carbon emitted during the construction stage.

Embodied carbon

Embodied carbon refers to the carbon dioxide emissions produced during the construction phase when materials are manufactured and transported to the site, and demolition and construction occur.

Cement and steel are responsible for most embodied carbon emissions in building materials, and 50 percent of global demand for cement and 30 percent for steel comes from the building and construction sectors.

The US National Trust for Historic Preservation found that it takes between 10 and 80 years for a new structure built with the latest energy efficiency measures to negate the carbon emissions resulting from its construction. Therefore, improving the efficiency of existing buildings is a key to lowering emissions.

⊕ **021**

Manufacturing and construction emissions

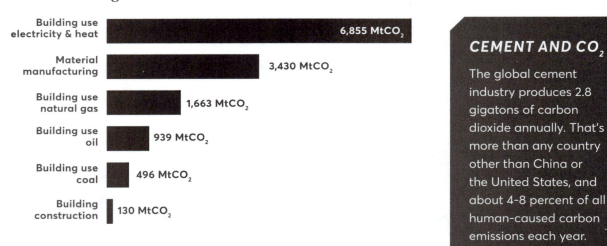

Category	Emissions
Building use electricity & heat	6,855 MtCO$_2$
Material manufacturing	3,430 MtCO$_2$
Building use natural gas	1,663 MtCO$_2$
Building use oil	939 MtCO$_2$
Building use coal	496 MtCO$_2$
Building construction	130 MtCO$_2$

CEMENT AND CO$_2$

The global cement industry produces 2.8 gigatons of carbon dioxide annually. That's more than any country other than China or the United States, and about 4-8 percent of all human-caused carbon emissions each year.

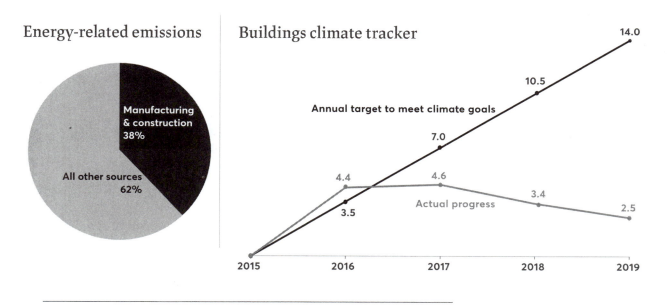

Energy-related emissions

Manufacturing & construction 38%

All other sources 62%

Buildings climate tracker

Annual target to meet climate goals

14.0
10.5
7.0
4.4
3.5
4.6
3.4
2.5

Actual progress

2015 2016 2017 2018 2019

What is the use of a house if you haven't got a tolerable planet to put it on?

— Henry David Thoreau

"She's been acting like that ever since she made her home energy efficient."

Agriculture and Meat Production's Role in Climate Change

For the majority of the planet's population, food is supplied by the industrial agriculture sector and delivered through the worldwide transportation network. A significant percentage of the planet's carbon emissions (more than 20 percent) are a direct result of food production, and in particular, the production of meat and dairy products.

Annually, food production creates 13.6 gigatons of greenhouse gases. The reasons for this can be traced to three main factors related to meat production:

- destruction of carbon-reducing forests and habitats to make way for pasture land
- plants grown with carbon-producing fertilizers
- cattle and sheep producing large amounts of methane as a byproduct of bacteria breaking down their food in the digestive process

Of the total emissions from food production, about 61 percent are due to livestock raised for meat and dairy. Collectively, livestock are one of the largest sources of methane emissions globally. As a greenhouse gas, methane is shorter lived but is more than 80 times more potent as a contributor to global warming than CO_2 while it is in the atmosphere.

BEEF IS A PROBLEM

The biggest agricultural source of greenhouse gas emissions is beef, generating 30 pounds of CO_2 for every 1 pound of meat produced.

The .115 gigatons of methane produced annually by the meat and dairy industry is the planet-warming equivalent of 3.5 gigatons of carbon, or the annual CO_2 emissions from the EU. Methane is also key in the formation of ozone and smog, leading to poorer air quality.

Carbon footprint of a cheeseburger

According to SixDegrees News, the carbon footprint of a cheeseburger is 4 kg of CO_2 equivalent gases. Of that amount, .5 kg is from diesel emissions, .9 kg is from electricity emissions, and 2.6 kg is from cattle methane emissions.

A gallon of gasoline emits 8.8 kg of CO_2, so a cheeseburger has the same emissions as about one-half a gallon of gasoline.

A single cheeseburger has the same climate impact as driving a typical car more than 10 miles/20 km.

⊕ 022

Urban Heat Islands

Cities are made with materials like asphalt and concrete that trap heat during the day, and radiate it back at night. How cities are laid out also affects whether this heat is trapped or is able to disperse.

As emissions and sunlight interact with already hot air in cities, it can create urban heat islands. This creates even more stagnant air that traps more pollution. When heat degrades airborne pollution, ground-level ozone is created. This ozone traps even more pollution, accelerating the greenhouse effect and making it even hotter while the air quality worsens.

This tends to drive people indoors, where they turn on the air conditioner, amplifying a vicious cycle. Not only do air conditioners consume a lot of energy, creating more emissions from power plants, but they also leak hydrofluorocarbons, which have a 1000x more negative effect on the environment than carbon dioxide.

This means that pollution in urban areas traps heat and makes pollution and global warming worse by increasing the demand for cooling, resulting in additional emissions of heat-trapping greenhouse gases. All of this creates a continuous cycle which leads to cities getting hotter while air quality gets worse.

The compounding effect of heat on pollution on urban areas is one of the reasons that the 1.5°C/2.7°F number is so important in understanding climate change and the urgency around reducing emissions. The more pollution we create, the hotter it gets and the more ecosystems are irreparably damaged.

⊕ **359**

Urban heat

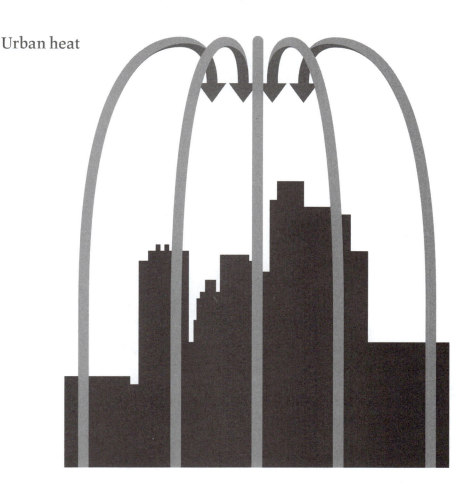

Cities trap heat due to pollution from factories and transportation, and materials such as concrete and asphalt.

And the air conditioning used to cool us down emits more heat and even HFCs.

Solutions include: more trees, lighter-colored roofs and roads and more efficient machinery. These solutions need to be distributed equitably.

The Plastic Lifecycle

In 2020, 367 million metric tons of plastic were manufactured worldwide. From food packaging to computer casings to clothing to water bottles, plastic is used every day. Worldwide plastic production is up more than 50 percent since the year 2000.

Throughout its life, 1 kg of plastic will produce approximately 6 kg of CO_2. This means that all of the plastic produced in 2020 will emit 2.2 megatons of CO_2 over the course of its lifetime.

Plastics emit carbon from start to finish. They are created with raw materials like oil and often incinerated when discarded.

Extraction

Most plastics are sourced from fossil fuels in the form of oil, natural gas, or coal. These non-renewable resources emit greenhouse gases when extracted from the earth. In 2018, petroleum and natural gas production for the US plastic and petrochemical industries alone emitted the equivalent of more than 72 million metric tons of carbon dioxide.

Manufacturing and use

Raw materials meant to be used as plastics are refined into plastic pellets, a process that emits more greenhouse gases. Plastic pellets are then manufactured into end-use products.

Up to 50 percent of plastics created are single-use: they're meant to be used once and then thrown away. This means that nearly half of all produced plastic is being created with the knowledge that it will end up in the trash very soon.

End of life

After a plastic product has been used, it's either thrown away or recycled. Since 1950, it's been estimated that globally only 9 percent of trashed plastics have been recycled, with over 79 percent going to landfills, incineration, or the natural environment.

At least 14 million tons of plastic end up in the ocean every year, and plastic makes up 80 percent of all marine debris found from surface waters to deep-sea sediments.

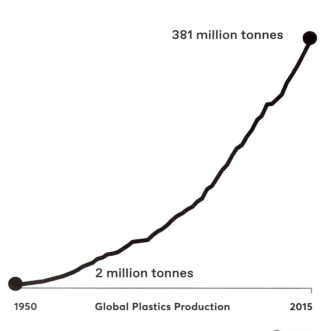

381 million tonnes

2 million tonnes

| 1950 | Global Plastics Production | 2015 |

🌐 **027**

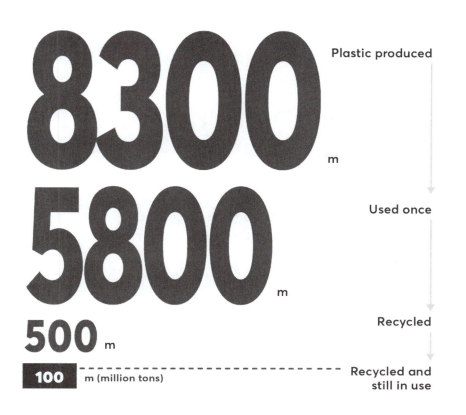

8300 m — Plastic produced

5800 m — Used once

500 m — Recycled

100 m (million tons) — Recycled and still in use

Plastic is rarely successfully recycled.

The True Cost of Plastic

Plastic is difficult to dispose of or recycle, but the majority of its emissions impact is created not at the end of its life cycle but at the beginning: 91 percent of plastic-related emissions are generated in the manufacturing process itself.

About half of all plastics are made from ethane, a chemical contained in natural gas. Fracking drills must be constructed to harvest ethane, resulting in deforestation of large areas: 19.2 million acres in the US alone have been cleared for extraction, causing an estimated 1.7 gigatons of carbon dioxide emissions and removing trees equivalent to a 6.5 million metric tons-per-year carbon sink. Once the drills are operational, extraction emits large amounts of carbon dioxide and methane.

After its extraction, ethane is transported to an ethane cracker where it is refined into ethylene. This process generates over 260 million tons of carbon dioxide emissions per year worldwide, or about 0.8 percent of the world's total emissions. Ethane crackers produce other pollutants as well, including carcinogens.

The ethylene molecules are then chemically combined to form long strands (polymers), creating polyethylene (PET) resin in the form of pellets. These pellets are sold and transported to the end-use manufacturing facilities where plastic items are made. It is estimated that up to 500 million tons of carbon dioxide worldwide is emitted yearly from polyethylene production. The total emissions of all plastics, including PVC, polystyrene, and HDPE, are estimated at 860 million tons.

🌐 **346**

Global Witness reports that Shell Oil's carbon capture project has produced more carbon than it has captured over the last five years.

The Dust Bowl: Lessons Learned for Farmers Everywhere

The Dust Bowl of the 1930s had a devastating environmental, economic, and social impact. It did, however, lead to advances in our understanding of three key concepts:

- enhanced soil and resource conservation practices
- government involvement in agricultural policy-making, along with farmland administration activities
- a deeper study of atmospheric and climatic activity, and the influence of such events on society and human migration

This particular disaster affected the Great Plains of North America for most of the 1930s. The afflicted areas included ten US states and three provinces in Canada. In addition to teaching us new ways to deal with certain climate disasters, it explains why 2.5 million people migrated to find employment elsewhere and escape starvation.

Several years of abnormally low rainfall and increased erosion from uncommonly frequent high-wind events fueled a series of huge dust storms. One such storm, originating on the Great Plains in May 1934, deposited 6 million tons of dust on Chicago in one day before moving to major East Coast cities from Washington to Boston.

Experts now agree that the desolate conditions had both natural and human-induced causes. The poor soil and resource management practices of tenant farmers exacerbated the effect of high winds and limited rainfall.

As the dust storms and droughts persisted into the mid-1930s, many crops in the region failed, farmers had little food to eat, and thousands lost their homes to foreclosure. Twenty percent of the population in the worst affected counties left the area. Many traveled to adjacent counties or states, but some went as far as California.

> *We are not made wise by the recollection of our past, but by the responsibility for our future.*
>
> — George Bernard Shaw

It wasn't just farmers who emigrated. A Bureau of Agricultural Economics study of the 116,000 families that arrived in California in the 1930s from southwestern states found that only 43 percent were farming before they migrated.

The Dust Bowl disaster and subsequent migration of millions of people left a lasting legacy. It led to expanded governmental participation in land and soil management administration. Agricultural and atmospheric scientists learned that practices such as soil and water conservation, crop rotation, and adequate wind protection must be managed by all farmers to protect and nurture the land for continued productive use.

🌐 **337**

Carbon Inequality, Climate Change, and Class

Wealth is directly correlated to the amount of carbon emissions generated yet *inversely* correlated to the estimated impact of the resulting climate change.

The rich produced the most carbon emissions from 1990 to 2015. At the same time, the poor worldwide are the most likely to be impacted. Some of the reasons are:

- A greater percentage of the poor work in agriculture, which is particularly vulnerable.
- With fewer household resources in reserve, a natural disaster is more likely to impact their food, water, and health.
- Drainage, sewage, flood control, and other forms of civic infrastructure in poor communities are often less developed.
- Many of the largest cities with impoverished residents are at or near sea level.
- Healthcare resources are more limited in poor communities.

🌐 **357**

Carbon inequality

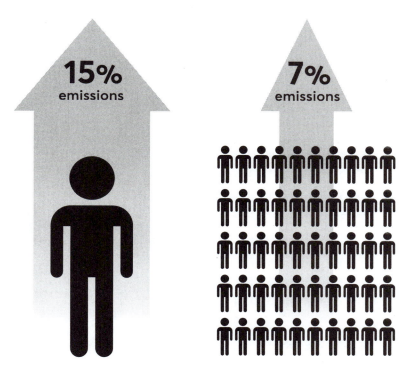

15% emissions

7% emissions

The richest 1 percent of humanity was responsible for 15 percent of global emissions compared to 7 percent from the poorest 50 percent. That's more than double the emissions from a small fraction of the people.

CHECK OUR WORK

The Almanac is based on thousands of sources. Don't take our word for it. Look for this number at the end of an article and then visit **www.thecarbonalmanac.org/000** (but replace 000 with your article number). **Dig deep and share what you learn.**

Paving the Planet

Fathom Information Design created a series of maps showing the topology of countries based on what they've paved. Every road on the map was built in the last two hundred years. The roads now define the place.
⊕ **375**

CO$_2$ Emissions from Global Shipping

Global shipping is vital for transporting bulk and containerized goods in the globalized economy. Essential raw materials like crude oil, iron ore, bauxite, and coal are carried in dedicated bulk carriers. Specialized shipping includes carrying cars, trucks, and heavy machinery. Refined fuels like diesel, gasoline, and liquefied natural gas are brought to markets in dedicated ships. Much of the world's manufactured goods are transported in shipping containers on large container ships.

The global shipping sector produced just over 1 gigaton of CO$_2$ in 2018 (2.51 percent of global emissions), a nearly 10 percent increase over 2012 levels. The International Maritime Organization (IMO) projects that shipping emissions will increase by up to 50 percent by 2050 under business as usual scenarios.

Commercial bulk shipping is one of the lowest GHG emissions forms of cargo transport available when measured per ton-km. However, the vast distances covered by international shipping offset this advantage.

The majority of emissions from commercial shipping are from the use of residual fuel oil, also known as resid, No. 6 fuel oil, heavy fuel oil, or bunker fuel. Bunker fuel is a low-value byproduct of oil refining. It is low-cost and energy dense, making it ideal for shipping companies wanting to cut costs. However, it creates more GHG than other fossil fuels. The IMO has screened a wide range of potential decarbonization strategies for shipping. These fall into several categories:

- **Energy-saving technologies:** Engine improvements, advancements in auxiliary systems like pumps and fans, incremental innovations in steam plants, and the reuse of waste heat, improvements in propeller design and maintenance, hull maintenance (to reduce water drag), and reducing ship weight.
- **Renewable energy usage:** Incorporating wind power (towing kite or sails) and solar photovoltaic panels.
- **Alternative fuels:** Hydrogen, ammonia, synthetic methane, biomethane, methanol, and ethanol.
- **Operational improvements:** Reduced operating speed to reduce drag.

Work has begun on the development of on-board CO$_2$ capture for ships. These systems would separate and purify CO$_2$ from combustion gases, with the captured CO$_2$ being compressed and liquefied. The stored CO$_2$ could be offloaded at the next port for geological storage.

🌐 **373**

TRANSPORT MODE	CO$_2$ (TON/KM)	CARGO TYPES
Bulk shipping	4.5	Bulk agriculture, forestry, minerals, coal
Oil tanker	5.0	Crude oil
Chemicals tanker	10.1	Chemical products
Container shipping	12.1	Most non-bulk goods types
Liquefied natural gas tanker	16.3	Natural gas
Rail	22.7	Various
Road	119.7	Various
US Air Cargo Fleet (average)	963.5	Various

FUEL TYPE	ENERGY DENSITY (GJ/M3)	COMBUSTION GHG EMISSIONS (KG CO$_2$e / GJ)
Bioethanol*	23.4	2.5
Liquefied natural gas	25.3	54.5
Gasoline	34.2	69.6
Diesel	38.6	70.4
Residual fuel oil	39.7	74.2

* Bioethanol produces CO$_2$ emissions when combusted. Because this fuel isn't pumped from underground, it's part of the fast cycle and isn't considered a net emission. This is why bioethanol combustion emissions are so low. Due to low energy density and relatively high cost per unit of energy, very few biofuels (0.1 percent by energy use) are used in international shipping.

The Domino Effect in Action: Urea

For years, urea has been added to the fuel used in modern diesel trucks, cars, and tractors. While urea occurs in nature (extracted from urine), mass-produced urea can be artificially synthesized using inorganic starting materials. This is often done in conjunction with manufacturing fertilizer.

Since 2010, urea has served to reduce the nitrous oxide emissions created by combustion engines in diesel trucks. About 220 million tons of urea are produced globally per year.

Supply chain issues and increased demand have created shortages that highlight how fragile the systems that provide urea are:

- Hurricane Ida's devastation on the US Gulf Coast in August 2021 caused suspension of key refineries, contributing to shortfalls of fertilizer. Heat and drought have reduced crop yields and added to food insecurity in vulnerable areas around the world.
- Rising natural gas prices first slowed urea production, and then China's electricity rationing put a greater strain on factory output. As a result, China, the world's largest producer, ceased exporting urea.
- Since the beginning of the COVID-19 pandemic, China decreased exports of urea.
- Family farmers in India have little to no available fertilizer for their fields. Some of these fields were already experiencing reduced crop yields due to weather-related effects.
- With limited supply, urea-based fertilizer now demands a higher price. Already at their highest since 2011, additional increases in price will add to food insecurity.
- The escalating cost of fossil fuels, particularly coal and natural gas, add to the cost of urea-based products as they are used in the production process.
- Some truck drivers in South Korea and Australia found their trucks disabled without the urea required to reduce greenhouse gases.
- The impact on farmers in India and truck drivers in South Korea and elsewhere have direct impacts on families' livelihoods and food security.

⊕ **376**

Don't forget nature. Because today, the destruction of nature accounts for more global emissions than all the cars and trucks in the world. We can put solar panels on every house, we can turn every car into an electric vehicle, but as long as Sumatra burns, we will have failed. So long as the Amazon's great forests are slashed and burned, so long as the protected lands of tribal people, indigenous people are allowed to be encroached upon, so long as wetlands and bog peats are destroyed, our climate goals will remain out of reach . . .

— Harrison Ford

Effects of Carbon-Based Cooking Fuel

Approximately 10 percent of the world's population has no access to electricity and 2.6 billion people are without safe fuel for cooking. As a result, these households rely on solid, carbon-based fuels (such as wood, crop wastes, charcoal, coal, dung, and kerosene) for cooking, as well as heating and lighting. These cookstoves present health and safety risks to the users and also emit a significant number of greenhouse gases.

Cookstoves produce an estimated 2.3 percent of global CO_2 emissions. India and China have the most cookstove users, and also have the biggest impact on climate. However, smaller countries like Azerbaijan and Ukraine also have a large impact on temperature, because their emissions have an impact on Arctic snow—which counteracts its reflective cooling effect.

🌐 606

Clean Fuels 12%

Polluting Fuels 88%

Cooking fuel sources in low-income countries

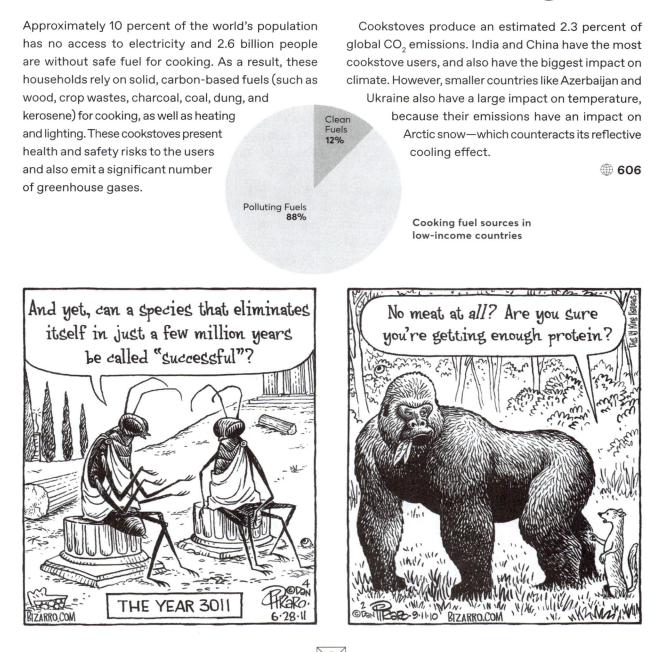

YOU CAN MAKE A DIFFERENCE

Visit **www.thecarbonalmanac.org** and sign up for **The Daily Difference**, a free email that will connect you with our community. Every day, you will join thousands of other people in connecting around specific actions and issues that will add up to a significant impact.

Everything costs and costs the earth. — Maya Angelou

Short-Lived Climate Pollutants

Air pollution leads to more than 6.7 million premature deaths each year. Short-lived climate pollutants (SLCPs) are a significant contributor.

These greenhouse gases and air pollutants degrade air quality and also have a near-term warming impact on climate. SLCPs include hydrofluorocarbons, black carbon, methane, and ozone at ground level. While soot (black carbon) can be seen in the air, the others are invisible.

SLCPs only last a short time in the atmosphere, ranging from a few weeks (soot) to twenty years (methane and HFCs). Once in the atmosphere, they decrease air quality and have a measurable effect on increasing the temperature of the atmosphere.

Because they act so quickly, reducing SLCPs immediately will reduce climate change in the near-term. Studies estimate that reducing SLCPs immediately could reduce climate change by 4°C/7.2°F and reduce predicted Arctic warming by 50 percent before 2050.

It's further argued that reducing SLCPs would slow the rate of sea-level rise by roughly 24-50 percent. In addition, reducing ozone has the potential to prevent the loss of more than 50 million tons of crops each year.

⊕ **363**

I used to think the top global environmental problems were biodiversity loss, ecosystem collapse, and climate change. I thought with 30 years of good science we could address those problems, but I was wrong.

The top environmental problems are selfishness, greed, and apathy—and to deal with these we need a spiritual and cultural transformation, and we scientists don't know how to do that.

— Gus Speth

Smoke Signals:
A Global Warning from Australia

The continent of Australia has an annual fire season, but, in 2008, the government was warned that change was coming. The Garnaut Review offered the following prediction: "Recent projections of fire weather suggest that fire seasons will start earlier, end slightly later, and generally be more intense. This effect increases over time but should be directly observable by 2020."

Projections were correct. 2019 was the hottest and driest year since Australia began keeping records in 1910. The 2019 fire season started earlier in September and ended later in March of 2020, and included some of the worst natural disasters in history, shocking the rest of the world. Megafires burned uncontrollably, creating weather systems: black hail, "firenadoes," and pyrocumulus clouds that produced lightning without rain, igniting even more fires. 3,400 homes and many additional buildings were burned. Mass evacuations saved thousands of lives, but not all: 33 people died, including nine firefighters (six Australian, three American).

Fires raged through 21 percent of Australia's forests including world heritage areas, national parks, and ancient rainforests that had never burned before. Australia's wildlife was irreversibly impacted by this widespread loss of habitat.

About three billion animals were in the path of the wildfires, including:
- 61,000 koalas
- 1 million wombats
- 5 million kangaroos and wallabies
- 5 million bats
- 39 million possums and gliders
- 50 million native rats and mice
- 143 million other native mammals
- 2.46 billion reptiles
- 100,000 cows and sheep

Health care costs resulting from smoke inhalation alone were estimated to have cost the Australian Government $1.95 billion AUD ($1.4 billion USD).

2019-2020 also saw more fires across the globe, including blazes in the Arctic, Amazon, Canada, Greenland, Indonesia, Russia, and the United States. Niklas Hagelberg, of the United Nations Environment Programme (UNEP), warned, "Megafires may well become the new normal as global temperatures continue to rise."

🌐 **343**

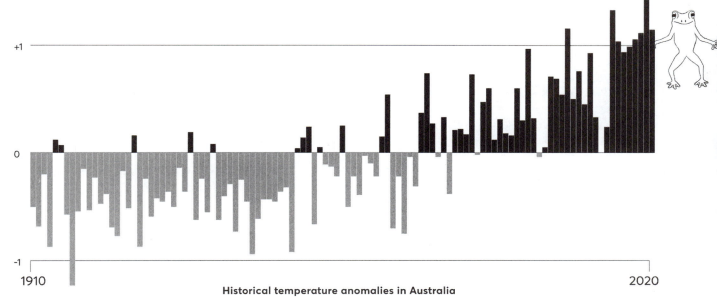

Historical temperature anomalies in Australia

OpenAI, a company that deploys artificial intelligence, consumed 1287 megawatt-hours to train its latest language model to learn human text—about the same amount 120 average American households used in 2020.

Computing and Carbon

Computers don't make a lot of noise or spew exhaust, but they all use electricity, and electricity generation often produces carbon.

The tables compare the electricity consumption of small-scale and large-scale digital activities and devices.

🌐 **340**

Small scale electricity consumption of computing activities

ACTIVITY	ELECTRICITY CONSUMED	EQUIVALENT TO	CO_2 EMISSIONS PRODUCED
One hour of smartphone charging	3.68 Wh	0.67 LED bulbs/hr	0.00313 pounds/kWh
One hour of laptop use	45 Wh	8.18 LED bulbs/hr	0.0383 pounds/kWh
One hour of Netflix streaming (average across all viewing devices in 2019)	77 Wh	14 LED bulbs/hr	0.0655 pounds/kWh
One hour of playing Fortnite on Xbox X	148 Wh	26.91 LED bulbs/hr	0.126 pounds/kWh
One hour of playing Fortnite on PlayStation 5	216 Wh	39.3 LED bulbs/hr	0.153 pounds/kWh
One hour of desktop computer use	330 Wh	60 LED bulbs/hr	0.28 pounds/kWh

Large scale electricity consumption of computing activities (in 2020)

ENTITY	ELECTRICITY CONSUMED	EQUIVALENT TO	CO_2 EMISSIONS PRODUCED
Bitcoin network	66.91 TWh	170% New Zealand's total consumption	56.873 billion pounds/kWh CO_2
Google global network	15.139 TWh	39% New Zealand's total consumption	12.868 billion pounds/kWh CO_2

Recycling Paper

In 2022, 416 million tons of paper will be produced. That number is expected to increase in coming years due to packing needs in the e-commerce sector. A move away from plastics as packaging material has also resulted in an increase in paper as a replacement.

Paper is recycled from three sources. **Mill broke** paper is the scrap collected at the paper mill. **Pre-consumer** waste is found at the printer or at warehouses, and **post-consumer** waste is found at homes. In order to recycle paper, the ink has to be removed first. This technology was invented by the lawyer Justus Claproth in Germany in the late 1700s.

Paper facts
- 40 percent of paper pulp comes from wood (wood cellulose).
- 35 percent of trees felled across the world are used for paper production.
- Recycling 1 ton of newsprint saves about 1 ton of wood.
- Recycling 1 ton of copier paper saves about 2 tons of wood.

The paper-carbon connection
Paper fiber contains carbon. As it breaks down, it releases methane into the atmosphere. Recycling the paper keeps the carbon locked in for longer. Virgin paper can be recycled five to six times before it becomes un-recyclable because of the shortness of its fibers. In the EU, over 70 percent of all paper waste is recycled. The US recycles about 68 percent. India recycles about 30 percent.

Recycled paper facts
- creates 74 percent less air pollution than making virgin paper
- 40 percent decline in energy needed to manufacture recycled paper
- reduces landfill paper waste
- causes 35 percent less water pollution

🌐 **372**

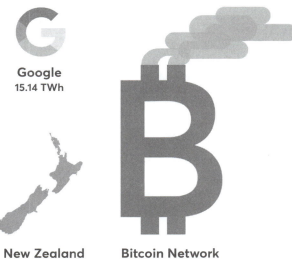

Google
15.14 TWh

New Zealand
39.36 TWh

Bitcoin Network
66.91 TWh

Trading Bitcoin uses far more electricity than New Zealand and four times more than Google worldwide.

IT WOULD TAKE...
- **1.5 billion** trees planted to offset the returns from online purchases made in the United States
- **1.6 billion** trees planted to offset emissions related to email spam
- **231 million** trees planted to offset emissions resulting from the data consumption of US citizens used in 2019
- **16 million** trees planted to offset the emissions resulting from the nearly 2 trillion annual Google searches

🌐 **340**

Soil now contains more than three times as much carbon as the atmosphere. It's estimated that this has decreased by more than 50% due to human activity, including farming.

The Climate Cost of Gas-Powered Leaf Blowers

Lawns cover 40 to 50 million acres in the United States. Residential property makes up 40 percent of that acreage. Lawns are common in Australia, Canada, and the U.K., but no other country comes close to the American obsession with having and maintaining their lawn.

There is a climate cost to maintaining lawns and one of the worst offenders is the gas-powered leaf blower. The CA Air Resources Board reports that gas-powered leaf blowers are bigger polluters than automobiles.

smog-causing hydrocarbons and nitrous oxides as well as carbon monoxide and particulate matter, all of which cause harm to human health. As a greenhouse gas, nitrous oxide is 300x more potent than CO_2.

More than 100 cities and towns in the US have banned gas-powered leaf blowers or limited their use. In many cases, people are switching from gas-powered to electric blowers. Though they are not perfect (they still require power plants to produce electricity) the impact is far lower.

1 hour leaf blower use

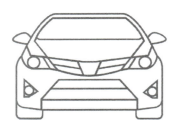

1770.278 kilometers driven in a Toyota Camry

The results of a 2011 study showed for every hour of use, a leaf blower emits 299 times as many carcinogenic hydrocarbons as a Ford F150 SVT Raptor pickup truck.

Most consumer-level leaf blowers use a two-stroke engine. These engines lack a separate lubricant system and must therefore mix fuel with oil. Approximately 30 percent of the engine's fuel never completes combustion and is emitted as toxic pollutants.

According to the New York State Department of Environmental Conservation, the exhaust created contains

Many horticulturists recommend foregoing blowers and rakes altogether and leaving the leaves where they fall. Fallen leaves will break down and improve the soil while also providing a winter haven for pollinators and insects eaten by birds and other wildlife.

🌐 **034**

Heating the Outdoors with Patio Heaters

A propane patio heater and a speeding truck put out equivalent emissions, but the patio heater lacks the mechanisms by which the truck filters or reduces its emissions.

Operating one patio heater on most evenings and some lunchtimes is projected to produce four tons of carbon dioxide per year. That's about two-thirds of the total carbon dioxide produced by an average household. Multiply that by the 6 to 12 propane patio heaters found running in the average restaurant's outdoor dining space, and you get a sense of their environmental impact.

Arising as an outdoor solution in response to the global pandemic, demand for propane patio heaters more than tripled in 2020. Globally, the 2020 market was 365.4 million and is projected to grow to 535.6 million by 2026.

In 2019, propane patio heaters accounted for almost 57 percent of the global heater market, followed by 36 percent for electric. On average, a gas or propane patio heater emits 3400kg of carbon dioxide per year while an electric heater uses the equivalent of 500kg of carbon dioxide.

In 2019, North America had the highest sales volume for patio heaters (49 percent), with Europe second (34 percent), and Asia-Pacific third (15 percent). These regions are expected to continue to dominate the market.

Some European countries are now regulating patio heaters. A nationwide ban in France was postponed due to Covid-19 but becomes operative in 2022. And some French cities like Lyon have already issued their own municipal directives against the outdoor heaters. A nationwide study estimated that patio heaters in France emit half a million tons of carbon dioxide per year.

🌐 **360**

Google search interest in "patio heater"
Relative interest 2017-2022

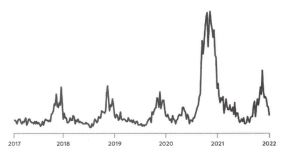

2017   2018   2019   2020   2021   2022

PATIO HEATERS

Also known as umbrella or mushroom heaters, patio heaters are radiant heat appliances that create thermal radiation for outdoor spaces within the vicinity of the heater. Most are fueled by natural or propane gas, though electric patio heaters are also available.

Scenarios

What's likely to occur if we
choose to act (or if we don't)?

The Five Scenarios Outlined by the IPCC

Predicting the future has never been more urgent. Thousands of climate scientists and economists have worked together to build and test rigorous computer models to estimate what the planet will be like one or two generations from now.

The **Intergovernmental Panel on Climate Change (IPCC)** consists of volunteer scientists from all over the world who evaluate and find consensus on current scientific knowledge regarding climate change—the past, present, and future risks and possibilities.

They've issued a series of reports, and created five likely outcomes for the world in 2050 and beyond. These scenarios are based on complex calculations measuring the climate's response to greenhouse gas emissions, land use, and air pollutants.

Possible future trajectories (or pathways) of economic growth, population, and greenhouse gas emissions are expected to cause rising average temperatures on Earth.

The scenarios' names are based on these Shared Socioeconomic Pathways (SSP) and are numbered 1 through 5—each a more negative outcome than the one before.

Five potential scenarios:

Warming occurs in *every* scenario, but there are substantial differences among them in regard to:

- intensity of weather
- sea rise
- heat waves
- loss of snow and ice
- action and policies moving forward

The scenarios illustrate how problems compound over time, and changes in current practices could have a large impact in years to come.

Previous estimates by the IPCC have proven to be too optimistic, so in their most recent IPCC report, the period in which global surface temperature is expected to cross the 1.5°C level is predicted to arrive 10 years earlier. Nevertheless, the data collected since the publication of that report shows more warming in the near term than what was aggressively estimated.

🌐 **033**

Carbon dioxide emissions (GtCO$_2$e/yr)

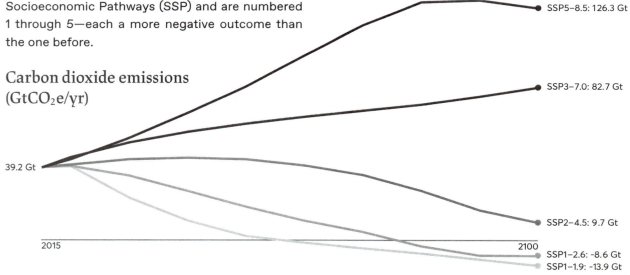

39.2 Gt

SSP5–8.5: 126.3 Gt

SSP3–7.0: 82.7 Gt

SSP2–4.5: 9.7 Gt

2015

2100

SSP1–2.6: -8.6 Gt

SSP1–1.9: -13.9 Gt

SCENARIO	°C/°F CHANGE	DESCRIPTION
#1 very low emissions (SSP1-1.9)	1.4°C / 2.5°F	Global CO_2 emissions are cut to net zero around 2050. This meets the Paris Agreement's goal of keeping global warming (at most) 1.5°C above preindustrial temperatures and then stabilizing around 1.4°C before 2100. Sustainable practices are adopted swiftly, shifting economic growth and investments. The effects of climate change are felt at a significantly lower intensity and rate than other scenarios.
#2 low emissions (SSP1-2.6)	1.8°C / 3.2°F	Global CO_2 emissions are still critically lowered, but insufficient to reach net zero by 2050. Temperatures stabilize around 1.8°C higher by end of 2100.
#3 intermediate emissions (SSP2-4.5)	2.7°C / 4.9°F	Progress towards sustainable practices is slow, similar to historic trends. CO_2 emissions stay at current levels. Net zero is not met by the end of the century. Temperatures rise by 2.7°C by 2100.
#4 high emissions (SSP3-7.0)	3.6°C / 6.5°F	Emissions and temperatures rise steadily, roughly doubling current levels. Countries shift toward competitiveness, more security, increased awareness of food supplies. Average temperatures have risen by 3.6°C by 2100.
#5 very high emissions (SSP5-8.5)	4.4°C / 7.9°F	CO_2 emissions are doubled by 2050. Increased energy consumption and the exploitation of fossil fuels powers economic growth, but... . The average global temperature rises 4.4°C by 2100.

IPCC scenarios

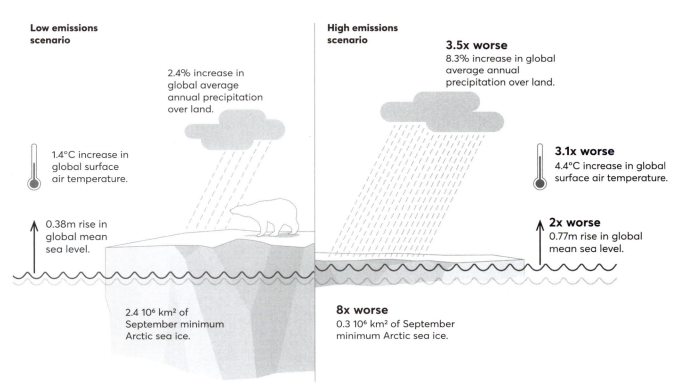

Low emissions scenario

2.4% increase in global average annual precipitation over land.

1.4°C increase in global surface air temperature.

0.38m rise in global mean sea level.

2.4 10^6 km² of September minimum Arctic sea ice.

High emissions scenario

3.5x worse
8.3% increase in global average annual precipitation over land.

3.1x worse
4.4°C increase in global surface air temperature.

2x worse
0.77m rise in global mean sea level.

8x worse
0.3 10^6 km² of September minimum Arctic sea ice.

Understanding the Five Scenarios

Visualizing the consequences of collective action is an essential step to moving forward. The IPCC scenarios clearly lay out what lies ahead.

Their report shows that the world has the scientific understanding, technical capacity, and financial means to limit emissions to 1.5°C/2.7°F, but also makes it clear that bold action and political will are imperative.

ON THE KIDS

Children aged 5 or less bear 90 percent of the burden of climate-related diseases.

Half a degree of warming makes a big difference

AREA OF CONCERN	SCENARIO 1	SCENARIO 2	DIFFERENCE
Global warming: Global mean surface temperature increase compared to preindustrial levels	1.5°C	2°C	0.5°C more
Severe heat waves: Global population exposed to severe heat waves at least once every 5 years	14%	37%	2.6x worse
Sea-level rise: Global population at risk per year from sea-level rise by 2100	69 million	79 million	10 million more
Sea ice levels: Frequency of ice-free summers in the Arctic sea	At least once every 100 years	At least once every 10 years	10x worse
Biodiversity loss (vertebrates): Vertebrates that lose at least half of their geographical range	4%	8%	2x worse
Biodiversity loss (plants): Plants that lose at least half of their geographical range	8%	16%	2x worse
Biodiversity loss (insects): Insects that lose at least half of their geographical range	6%	18%	3x worse
Ecosystem transformations: Global land area affected by ecosystem transformations	7%	13%	1.9x worse
Coral reef loss: Loss of reef-building corals compared to today	70 to 90%	99%	1.2x worse
Crop yield decline: Global population exposed to decreases in crop yield	35 million	362 million	10.3x worse

Scenario 1

This is the only scenario that meets the Paris Agreement's goal of keeping global warming to 1.5°C/2.7°F above preindustrial temperatures.

In this scenario, extreme weather is more common, but the world has avoided the worst impacts of climate change. There will still be risks to health and changes to the climate, but they will be significantly less than other scenarios. Limiting global warming to 1.5*C/2.7*F will, however, require unprecedented transitions in energy, land, infrastructure, transportation, industrial systems, and more.

Limiting global warming to 1.5°C/2.7°F will require unprecedented transitions in energy, land, infrastructure, transportation, and industrial systems.

Scenario 2

In the low emissions scenario, the world breaches 1.5°C/2.7°F soon after 2030 but manages to achieve the Paris Agreement goal of staying below a 2°C/3.6°F increase above preindustrial levels by 2100.

Global CO_2 and non-CO_2 greenhouse gas emissions are cut as severely as in Scenario 1, but not as fast, only reaching net-zero emissions after 2050. As in Scenario 1, carbon dioxide will also need to be removed from the atmosphere via reforestation, carbon capture, and other methods.

Half a degree warmer may not seem like much of a difference. However, the IPCC report is clear that every additional half a degree will result in a significant increase in negative impacts on human and natural systems.

For example, extreme hot weather events—such as heat waves, fires, floods, and droughts—will become more intense and more frequent, sometimes occurring simultaneously. This, combined with sea-level rises and increases in ocean acidity, will lead not only to loss of habitats for humans and other species, but also a loss of food availability through reduced crop yields and fishery outputs. The IPCC estimates that up to several hundred million more people will be negatively impacted by climate-related risks in this scenario compared to Scenario 1.

Scenario 3

This is a scenario that assumes political and economic forces make it difficult to take quick and dramatic action in the short run.

With cumulative CO_2 emissions having a near-linear relationship with rising global surface temperature, the 1.5°C/2.7°F limit would likely be exceeded in the early 2030s, less than one decade from when this Almanac was published.

GHG emissions do not decline until 2050 in this scenario and as a result, by the end of the century, warming is expected to be around 2.7°C/4.9°F.

The last time temperatures were 2.5°C/4.5°F higher than preindustrial levels was estimated to be over 3 million years ago.

There will be regional variations in warming. Average warming will be higher over land than the ocean, and greater in the higher latitudes of the northern hemisphere than the southern. Compared to Antarctica, the Arctic is more sensitive to warming, and since the industrial era it has warmed two times faster than other parts of the world.

Precipitation will increase. In all scenarios where global warming exceeds 1.5°C/2.7°F, an increase in precipitation is expected, particularly over land. A 1-3 percent increase in mean precipitation (both global and annual) is expected per 1°C/1.8°F rise in global average surface temperature.

While overall precipitation rises, there will be regional differences due to latitude. Precipitation will increase in high latitudes and wet tropics, but decrease over dry zones, including parts of the sub-tropics, such as the Mediterranean, southern Africa, parts of Australia, and South America.

Precipitation will increase in high latitudes and wet tropics, but decrease over dry zones.

The Arctic sea ice will melt. Any scenario where there is a rise above 1.5°C/2.7°F shows increased likelihood that there will be practically no Arctic sea ice in September by the end of the century. This likelihood becomes a near certainty when warming reaches 2°C/3.6°F.

A rise in global surface temperature will result in a greater loss of mass from glaciers and ice sheets, leading to a rise in global mean sea levels (GMSL), which is expected to accelerate throughout the 21st century in the last three scenarios. The ocean will also become more acidic in these scenarios as the ocean absorbs more carbon due to rising emissions. Some systems will be irreversibly changed. Sustained global warming is likely to permanently cause:

- sea-level rise
- loss of ice sheets
- permafrost carbon release

Scenario 4

This scenario supposes that as global climate change worsens, international coordination will falter. Instead of combined efforts to address the problem, countries will look inward to focus on national interests, mainly around energy and food security.

With a high reliance on fossil fuels to deal with short-term emergencies, GHG emissions grow steadily. By 2100, CO_2 emissions nearly double, exceeding 80 gigatons per year. Warming is exacerbated by weak air pollution control and a steady rise in non-CO_2 emissions.

Temperatures soar. As countries fail to meet climate pledges, temperatures will likely rise 2°C/3.6°F in the 21st century, with the 1.5°C/2.7°F threshold likely to be crossed in less than a decade.

Precipitation and drought are amplified. In scenarios where global warming exceeds 2°C/3.6°F (scenarios 4 and 5), global average precipitation increases by 2.6 percent, compared to 1995-2014.

The oceans change. The global sea surface temperature rises 2.2°C/3.9°F by the end of the century. Rising ocean temperatures may affect the Atlantic Meridional Overturning Circulation (AMOC), the largest system of ocean currents. If the AMOC were to shut down, there would be far-reaching effects, such as shifting of monsoons and a decrease in precipitation over Europe and North America. The shutdown could be permanent.

If the AMOC were to shut down, there would be far-reaching effects, such as shifting of monsoons and a decrease in precipitation over Europe and North America. The shutdown could be permanent.

Rising ocean temperatures lead to a rise in GMSL due predominantly to thermal expansion. In any scenario where temperatures cross the 2°C/3.6°F mark, there is an increased likelihood of the collapse of the Antarctic ice sheet. This causes the GMSL to rise to at least 1 meter around 2100, with some predictions putting that number over 2 meters.

Scenario 5

In the face of a worsening climate emergency, this scenario envisions ever more intensive fossil fuel development and energy use. This leads to a significant increase in GHG emissions. Annual CO_2 emissions *double* before 2050 and exceed 120 gigatons before the end of the century.

Improvements in the technology of renewable energy, combined with their growing acceptance, make this scenario unlikely. However, carbon-cycle feedbacks could impact atmospheric concentrations, which could create a cycle of planetary response that leads here. And in light of the fact that global surface temperature is expected to cross the 1.5°C/2.7°F level in 10 years and warming in the near term is more than what was estimated, less likely scenarios should not be discounted.

In this scenario, a 1.5°C/2.7°F increase is considered very likely in the near term period, around 2027. A 2°C/3.6°F increase is likely to be reached within several decades, and by the end of the century a previously unimagined increase of 4.4°C/7.9°F could occur. Humans have never lived under such climate conditions.

Unlike other scenarios, this scenario assumes a strong level of air pollution control and projects a decline of "ozone precursors" except for methane in the mid to long term. Methane is projected to rise until 2070.

As in other scenarios, greater levels of warming are expected to amplify differences in regional warming trends. For example, compared to 1995-2014 temperature ranges, parts of Amazonia or other tropical land regions could rise by 8°C/14.4°F and other tropical land regions could increase by 6°C/10.8°F.

Precipitation dramatically increases. With higher levels of warming, it's expected that there will be greater amplification of high and low precipitation. Ice sheets will disappear and sea levels and temperatures will rise. The loss of the largest ice sheets in the world in Greenland and the Antarctic will contribute to sea-level rise, as well as the loss of glaciers. And because ice sheets grow slowly but melt quickly, any loss of mass may be irreversible.

As the oceans absorb more heat and become warmer, water expands. A sea-level rise of nearly one meter could impact the lives of nearly a billion people living in coastal regions, islands, and areas currently prone to flooding.

Search the web, plant a tree.

Visit **www.thecarbonalmanac.org/search** to install a simple extension that plants a tree every time you do some web searches.

A sea-level rise of nearly one meter could impact the lives of nearly a billion people living in coastal regions, islands, and areas currently prone to flooding.

We don't throw anything away. We just take our problems and turn them into somebody else's problems.

— Simon Sinek

⊕ 039

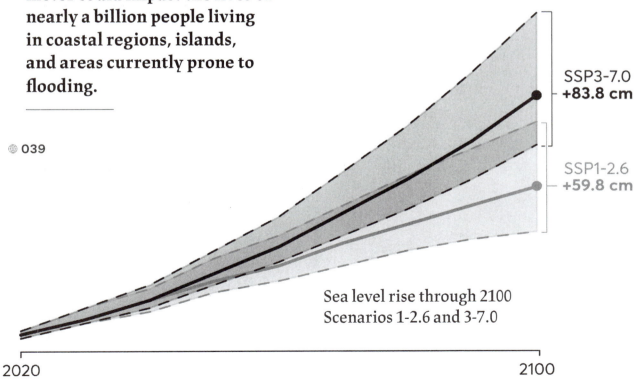

SSP3-7.0
+83.8 cm

SSP1-2.6
+59.8 cm

Sea level rise through 2100
Scenarios 1-2.6 and 3-7.0

2020

2100

10-, 50-, 100-, and 1,000-Year Climate Events

What are the chances?

One way to describe risk when it comes to climate is to talk about how often to expect a certain event to occur. The high-water mark on a dam, for example, could be the level a reservoir is likely to reach on average every ten years.

Climate change has upended these estimates. A once-in-ten-years heat wave might be likely to happen *nine* times in ten years under a scenario where temperatures increase by 4 degrees Celsius.

In the 50 years from 1970-2019, there was on average one disaster related to climate, weather, or a water hazard every day. Droughts, storms, floods, and extreme temperatures were among the top ten types of disasters in terms of human deaths and economic loss. According to the Secretary-General of the World Meteorological Society, disasters are becoming more frequent and more intense due to climate change.

Ten-year events

The IPCC examines three types of such events historically likely to happen once every ten years:

1. Hot temperature extremes over land
2. Heavy one-day precipitation over land
3. Agricultural and ecological droughts over dry regions

Under four emissions scenarios (increases of 1°C/1.8°F, 1.5°C/2.7°F, 2°C/3.6°F, and 4°C/7.2°F), these events would likely occur more frequently within a ten-year timeframe.

50-Year Events

The IPCC also examines hot temperature extremes over land historically happening once in 50 years. Extreme heat waves that historically happened every 50 years are now expected to happen close to every ten years under low- to mid-emission scenarios. The intensity of these events could be up to 5.3°C/9.54°F hotter in the 4°C/7.2°F emissions scenario.

Increasing frequency of 10-year heat waves

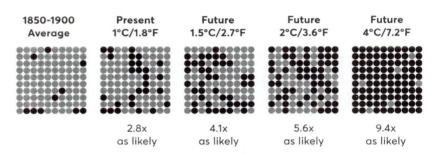

1850-1900 Average	Present 1°C/1.8°F	Future 1.5°C/2.7°F	Future 2°C/3.6°F	Future 4°C/7.2°F
	2.8x as likely	4.1x as likely	5.6x as likely	9.4x as likely

Heat waves could happen almost every year in a high emissions scenario. The intensity of the heat events could be up to 5.1°C/9.18°F hotter in the 4°C/7.2°F global warming scenario.

Increasing frequency of 10-year precipitation events

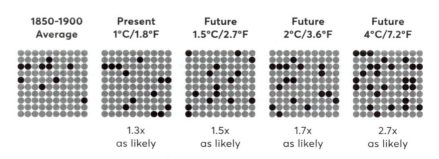

1850-1900 Average	Present 1°C/1.8°F	Future 1.5°C/2.7°F	Future 2°C/3.6°F	Future 4°C/7.2°F
	1.3x as likely	1.5x as likely	1.7x as likely	2.7x as likely

Precipitation events could be up to 30.2% wetter in the 4°C/7.2°F emissions scenario.

> *Plan for what is difficult while it is easy, do what is great while it is small...The worst calamities that befall an army arise from hesitation.*
>
> — Sun-Tzu

Increasing frequency of 10-year droughts

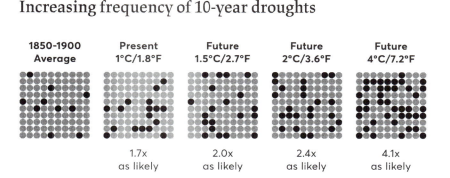

1850-1900 Average	Present 1°C/1.8°F	Future 1.5°C/2.7°F	Future 2°C/3.6°F	Future 4°C/7.2°F
	1.7x as likely	2.0x as likely	2.4x as likely	4.1x as likely

Drought events could be more intense by one standard deviation of soil moisture in the 4°C/7.2°F emissions scenario. Even in the 2°C/3.6°F emissions scenario it's possible that multiple climate events such as drought and heat are more likely to occur simultaneously, increasing the risks of other climate-related emergencies such as forest fire, food insecurity, and water quality.

Increasing frequency of 50-year heat waves

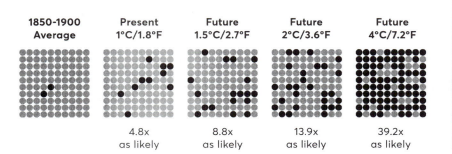

1850-1900 Average	Present 1°C/1.8°F	Future 1.5°C/2.7°F	Future 2°C/3.6°F	Future 4°C/7.2°F
	4.8x as likely	8.8x as likely	13.9x as likely	39.2x as likely

100- and 1,000-Year Events

Rainfalls that could be considered "once in a century" have happened several times in recent years in Australia, causing significant flooding. And the rainfall that happened in China during the summer of 2021 was considered a one-in-1,000-years flood. 617 mm of rain fell over three days, almost as much as the annual average. Other 1,000 year floods have also happened in Germany (2021) and in the USA during Hurricane Florence (2018) and Harvey (2017) and Tropical Storm Imelda (2019).

🌐 **680**

NORWAY GOES ELECTRIC

In 2021, only 8 percent of the new cars sold in Norway were powered by conventional gas or diesel engines.

Shifts in the Atlantic Ocean's Currents

The Atlantic overturning

We often think of the oceans as really big lakes. But it's more accurate to think of oceans as a series of giant intersecting rivers. The Atlantic Ocean's rivers, or currents, are known as the Atlantic Meridional Overturning Circulation (AMOC)—a phenomenon that is likely to be disrupted in scenarios 4 and 5.

The Gulf Stream is part of the AMOC. It carries more than 100 times the volume of water every second as the Amazon River.

The Gulf Stream brings warm water from the Caribbean north along the eastern coast of North America before crossing the Atlantic Ocean. Here it splits.

One part heads toward Greenland and the British Isles while the other circles down along the West coast of Africa. As the nutrient-rich warm water pushes north along the ocean surface, cold water flows south off the eastern coasts of North and South America.

Ireland is located at the same latitude as polar bear habitats in Canada. Ireland would be much, much colder without the warming waters of the Gulf Stream.

For thousands of years, this stable system has been responsible for the temperate climates of Western Europe and the eastern coast of North America. Now, that might be changing.

Today's slowdown, tomorrow's collapse?

Scientists have recently modeled the flow rate of the Gulf Stream for the past 2,000 years. They observe a 15 percent slowdown in the AMOC in the last 160 years. This slowdown is already causing climate disruption.

Most of the observed slowdown has occurred in the past 50 years and is linked to warming ocean temperatures caused by increased CO_2 emissions and melting glaciers. Scientists predict that the AMOC could slow by as much as 45 percent in the coming century. There is concern that it may collapse irreversibly, leading to significant climate disruptions throughout the world.

Rising sea levels

A sluggish Gulf Stream will make equatorial waters warmer and raise sea levels. The warm water will build up along the east coast of North America instead of traveling across the Atlantic to Europe.

For 15 months beginning in 2009, sea levels from New York to Newfoundland rose four inches

North America

So
An

Warm current from the Gulf of Mexico flows northeast across the Atlantic

Current cools and increases in density in the north, sinks to lower depths, and flows back south

Warm current from the Indian Ocean

due to a 30 percent slowdown in the circulation of the AMOC. In the Gulf of Maine, the ocean temperature has risen dramatically over the past decade. The region's cod fisheries have declined by 40 percent because of these changes in ocean temperature. At the same time, the warmer surface temperatures at the equator mean that hurricanes are becoming more common and more intense.

If the AMOC collapses, the entire east coast of North America would be affected by substantial sea-level rise. Millions of people would be displaced, and critical habitats for marine life—from commercially-fished species to endangered species like sea turtles and manatees—would be destroyed.

Melting glaciers

Melting glaciers from the Greenland ice sheet—the world's largest ice sheet outside of Antarctica—have created what scientists call the "cold blob" in the middle of the Atlantic Ocean. The cold blob is theorized to be both a symptom and a cause of the slowing Gulf Stream.

The cold blob in the Atlantic is also affecting weather. Scientists attribute bitterly cold winters, summer heat waves, and droughts to the changes in surface water temperature and humidity from the cold blob.

Geological records show that the last abrupt change to the AMOC caused severe droughts across northern Africa and plunged coastal regions surrounding the Atlantic back into Ice Age temperatures for 1,000 years.

🌐 **683**

Who Suffers Most?

The faster our climate warms, the greater the effects of extreme weather scenarios such as droughts, floods, heat waves, rising sea levels, and ocean acidity on people's livelihoods and homes. The degree of impact will mainly depend on each country's ability to implement adaptation and mitigation plans.

A country's ability to respond largely depends on its level of economic wealth. While climate change affects everyone, those who are most impacted will likely be those who have done the least to contribute to it and can least afford to mitigate its impact.

Most countries in the Northern Hemisphere (13 out of 15) that have the highest GDPs also produce the most GHG emissions per capita (PC) when compared to countries in the Southern Hemisphere.

For example, compare Bangladesh and the USA.

	POPULATION	HOUSEHOLD INCOME (2018)	CARBON EMISSIONS (PC)
Bangladesh	160,000,000	$1,698	0.5
United States	327,000,000	$63,062	15.2

Bangladesh is half the size in population, yet contributes less than 4 percent as much carbon per capita and has a per-person income that's less than 3 percent the size of the United States.

Due to its low elevation, Bangladesh is particularly affected by rising sea levels. By 2050, it's estimated that 18 million Bangladeshis will be displaced from their homes. Land previously used for crops is being claimed by rising sea levels, the increasing salinity of the soil harms crops and drinking water, and homes are regularly destroyed by tropical storms or the erosion of river banks.

Families who are still able to live in their homes are spending more money on repairing or preventing damage. With sea-level rise expected to continue beyond 2100 even if warming is limited to 1.5°C/2.7°F, Bangladeshis will continue to be disproportionately affected.

The IPCC identified the following populations as being at a "disproportionately higher risk of adverse consequences with global warming of 1.5°C/2.7°F and beyond":

- disadvantaged and vulnerable populations
- some Indigenous peoples
- local communities dependent on agricultural and coastal livelihoods

The regions identified as being at a disproportionately higher risk are:

- Arctic ecosystems
- dryland regions
- small island developing states
- least developed countries

Women have also been identified as particularly vulnerable to climate change by the UN. Seventy percent of the 1.3 billion people in poverty are women. Food and water scarcity particularly affect women due to their caretaking and farming responsibilities.

681

Don't take our word for it.

Visit **www.thecarbonalmanac.org/681** to check out this article's sources, relevant links, and updates.

Dig deep and share what you learn.

Ocean Acidity

Historically, the oceans have always had a basic pH of around 8.2. Due to the increase in CO_2 emissions, their pH is decreasing. Since pH is a logarithmic scale, with each unit decrease of pH, the oceans get 10x more acidic. As shown in the graph below, different future scenarios can have very different impacts on the oceans' acidity. The first two scenarios show a slight return to current levels in the long term, while the other three scenarios have a continuous downwards trend.

A democracy that doesn't include the voice of the environment will fail. We need the environment for co-existence.

— Oladesu Adenike

How are the oceans becoming more acidic?

The oceans absorb around one-third of all human CO_2 emissions. This CO_2 reacts with water and creates carbonic acid, which changes the acidity of the water in the ocean. More CO_2 equals more acidity.

Implications of a more acidic ocean

The first implication of a more acidic ocean is the decrease in the formation of corals and clams. The carbonic acid from the CO_2 emissions competes with corals and clams to get carbonate—a compound used to form their skeletons and shells.

Coral and shellfish are essential parts of the ocean ecosystem, and when they begin to suffer, so do the creatures that depend on them. Other marine lives can also be impacted since their respiration, calcification, photosynthesis, and reproduction are sensitive to pH variations.

⊕ **679**

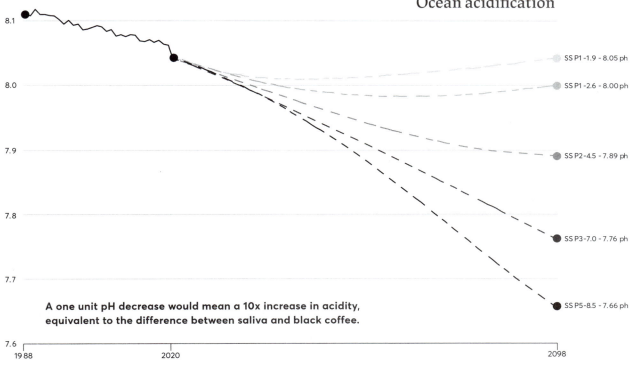

Ocean acidification

- SS P1 -1.9 - 8.05 ph
- SS P1 -2.6 - 8.00 ph
- SS P2 -4.5 - 7.89 ph
- SS P3 -7.0 - 7.76 ph
- SS P5 -8.5 - 7.66 ph

A one unit pH decrease would mean a 10x increase in acidity, equivalent to the difference between saliva and black coffee.

1988   2020   2098

Impacts

Climate impacts
everything around us

Threats to Coastal Communities

Over 40 percent of people on Earth live within 60 miles of a seacoast, including residents of eight out of ten of the largest cities in the world. Populations in many of these shoreline communities continue to grow even as the impacts of climate change become more visible and significant. This blend of climatic and geographic change in areas with high density means that residents of coastal communities are going to experience some of the earliest and most damaging effects of climate change.

Powerful hurricanes, increased rainfall, and shorter and warmer winters increase the likelihood and impact of flooding. This flooding can accelerate saltwater intrusion into groundwater supplies, which contaminates drinking water and irrigation for agriculture while also presenting an existential danger to ecological resources and aquatic ecosystems that are sensitive to increased salinization.

Shipping represents 80 percent of international trade and creates an estimated $14 trillion in annual global revenue. Rising sea levels also threaten ports and make critical infrastructure vulnerable.

Fisheries are endangered by ocean acidification, warming waters, and coral bleaching, and polluted runoff can feed toxic algae blooms that produce mass fish kills and create dead zones near the coast. Other industries that are central to coastal communities, like tourism, are also at risk.

🌐 601

Some places are especially vulnerable to rising sea levels and the effects of climate change.

These cities are some that face significant challenges.

TOKYO

MUMBAI

NEW YORK CITY

SHANGHAI

LOS ANGELES

CALCUTTA

BUENOS AIRES

LAGOS

BANGKOK

VENICE

BASRA

JAKARTA

ROTTERDAM

HO CHI MINH

Population Growth

One of the driving forces in climate change has been the growing population. There were 1.5 billion people on Earth in 1900, and there are more than five times that many now.

However, most countries' fertility rate (the number of children a woman can expect to have during childbearing years) has decreased since the late fifties. It is expected to fall to less than 2.1 children per woman by the end of the century—the fertility rate needed for a population to remain constant.

The world's population is projected to level out at 10.9 billion around 2100. Most of this growth will happen in Sub-Saharan Africa (with projections varying from 2.6 billion to 3.8 billion).

A second significant trend is the aging of the global population. The median age (the age where there's the same number of people above and below that line) will increase from 31 years in 2020 to 38 in 2050. The population older than 70 will increase sharply from 6 percent to 17 percent in this time frame. The speed and scale of these shifts are unprecedented.

Determining the impact of these figures on climate change is difficult. Feeding a larger population will indeed require more resources. However, most of the world's population growth will occur in countries that currently have a low impact on climate change.

⊕ **581**

World fertility rate (live births per woman)

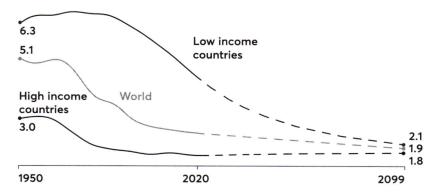

Projected worldwide population growth

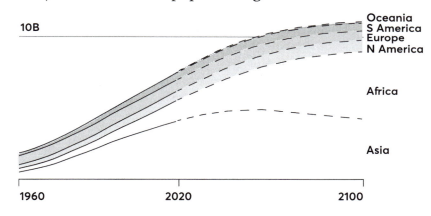

To me, the real challenge is the human mind, which is driving our actions: our beliefs and values shape the way we see the world, which in turn determines how wwe will treat it. So long as we assume that we are the center of the universe and everything revolves around us, we will not be able to see the dangers we create. To see those, we have to recognize that our very lives and our well-being depend on the richness of nature.

— David Suzuki

Human Migration Away from Inhospitable Land

Climate change is making many areas of the Earth increasingly unlivable. Climate stressors such as too little or too much rainfall, extended heat waves and droughts, and rising sea levels are forcing people to leave their homes and livelihoods behind.

Changing climates already result in climate migration within countries and across national borders. Statistics from the Internal Displacement Monitoring Center (IDMC) show that on average 22.7 million people are displaced annually by climate-related events (displacements include those caused by geophysical events such as earthquakes and volcanic eruptions). An Australian think-tank, Institute for Economics and Peace (IEP), reports that more than 1 billion people could face forced migration by 2050 due to climate change and conflict.

The majority of environmentally-induced migrants come from rural areas where their livelihoods depend on climate-sensitive sectors such as agriculture and fishing. Significant changes to growing conditions and seasons have lowered the dependability of crop yields and incomes for farmers. Similarly, changes to sea levels, water conditions, and other factors directly contribute to depleted fish stocks.

Climate migration out of urban areas is also happening in regions where sea-level rise affects densely populated coastal areas. Increased urbanization and sprawl are creating more heat waves and night-time temperatures no longer cool the daytime highs. Flooding will also increase from intensified rainfall events. All of these factors will drive further migration out of coastal cities. Over 10 million people were displaced by climate disasters over the six months between September 2020 and February 2021. Some 60 percent of these were in Asia.

More prosperous countries are already closing their borders to climate immigrants and places that are willing and able to accommodate migrants are becoming scarcer.

🌐 **068**

Weather-related human displacement in 2020

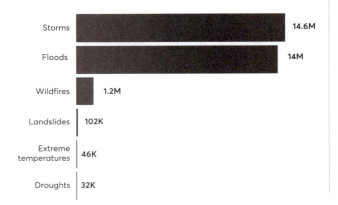

Storms	14.6M
Floods	14M
Wildfires	1.2M
Landslides	102K
Extreme temperatures	46K
Droughts	32K

Five countries with the most new displacements by disasters in 2019

India	5.0M
Philippines	4.1M
Bangladesh	4.1M
China	4.0M
United States	0.9M

The Impact of Climate Change on Indigenous Peoples

> *We all breathe this one air, we all drink the same water. We all live on this one planet. We need to protect the Earth. If we don't, the big winds will come and destroy the forest. Then you will feel the fear that we feel.*

— Raoni Metuktire

Despite contributing the least to greenhouse gas emissions, Indigenous peoples are on the front lines of effects from climate change. Their health as well as their livelihood is under threat from the changes occurring around the world.

Regardless of the location of their homelands, an estimated 370 million Indigenous people living in at least 90 countries are facing disproportionate threats due to:

- ancestral traditions and close connections to the natural world
- greater dependency on wildlife and natural resources for food and livelihood
- history of political & systemic marginalization
- greater likelihood of living in conditions of socio-economic distress
- greater likelihood of disease
- less access to quality healthcare

Tribal air, food, and water are affected by increasing temperatures and other consequences of increased emissions. A decrease in clean air and water as well as food availability for Indigenous peoples are directly tied to:

- the rise of sea levels
- pollution of surface waters
- decrease in snowpacks
- increase in wildfires
- extended droughts
- an increase in infestations and infectious diseases

Traditional lands are a fundamental factor in the health and well-being of Indigenous peoples. Impact on these lands, whether from climate change directly or the actions taken by governments to mitigate climate change effects, can negatively affect the health and well-being of Indigenous peoples. Beyond the effects on traditional food and natural medicines, the ability to perform traditional ceremonies is also impacted.

From the Arctic tundra and the Amazon Basin to the savannas of Africa and the island and coastal regions of the Pacific, forced climate migration or displacement from their traditional homelands is a traumatic experience for many Indigenous peoples who lived off that land for generations. Impacts to these communities from climate change are cultural as well as environmental.

🌐 595

BY THE SEA

Human activity in coastal regions on land and in the ocean accounts for nearly two-thirds of global GDP, but runoff, pollution, and overfishing put many of these resources at risk, while sea level rise and storm surges threaten cities and other infrastructure near the ocean.

Race, Fairness, and Climate

Extraction economies and industrialization have a long history related to colonial dominance and slavery. Every conversation about the impacts of climate change must also acknowledge the dynamics of class, race, and caste. The theft of lands, the wholesale clearing of forests, and the hierarchy of industrial work all are factors in producing the climate related problems we face now.

In the United States today, 56 percent of those who live near toxic waste sites are people of color. That's nearly twice as many as would be expected based on overall population percentages. And a 2014 study reports that non-white Americans were exposed to 38 percent higher levels of nitrogen dioxide than the white population.

Data shows that the wealthiest 10 percent of people in the United States emit more than four times as much in greenhouse gases as the average American, and yet the impacts of climate and pollution are far more likely to have an impact on the poor.

Worldwide, the countries that have traditionally been industrial and colonial powers have contributed the lion's share of carbon to the environment over time, while the poor are much more likely to bear the brunt of climate changes.

When climate disasters hit, it's usually the poorest who are the most likely to be affected, and the resources spent for rebuilding are rarely allocated evenly.

As countries move to invest in new technologies and resilient infrastructure, groups like the Climate Justice Alliance are arguing that—once again—the impacts and investments are not being distributed with justice and fairness in mind. They urge a shift from an extraction economy based on industrialization and scarcity to a regenerative one that is built on equity and abundance. The values filter they propose includes:

- Shifting economic control to communities
- Democratizing wealth and the workplace
- Advancing ecological restoration
- Driving racial justice and social equity
- Relocating most production and consumption
- Restoring cultures and traditions

⊕ **584**

I think about people who got the worst food, the worst health care, the worst treatment, and then when freed, were given lands that were eventually surrounded by things like petrochemical industries.

— Elizabeth Yeampierre

Displaced Human Communities

Climate change is on track to create the largest wave of human migration in history. It is estimated that the impacts of climate change on agricultural productivity, water resources, and social instability could drive over 200 million people to become climate refugees or be internally displaced by 2050.

While most greenhouse gases that cause climate change are released into the atmosphere by countries in the global north, populations in the global south are experiencing the greatest impacts, including events such as widespread crop failures, cycles of flooding and drought, and famine. These communities are also most likely to experience the greatest impact in the future. Some estimates say that by 2100 it could be too hot for people to even stand outside for more than a couple of hours in several climatic zones.

As sea levels rise, many island communities will become imperiled. Between 11 and 15 percent of small islands in the Pacific and Indian Oceans have elevations of under five meters. A sea-level rise of just one-half meter could displace 1.2 million people from low elevation islands. A rise of two meters could drive the displacement of 21.5 million people.

Climate change may also contribute to social disruption and government instability that drives displacement. The region called the Fertile Crescent saw an unprecedented drought in 2007-2010 that led to massive urban migration and unemployment. This contributed to the Arab Spring and the many crises that followed, including the war in Syria.

⊕ 602

Millions of people displaced by climate change

◯ **Millions**

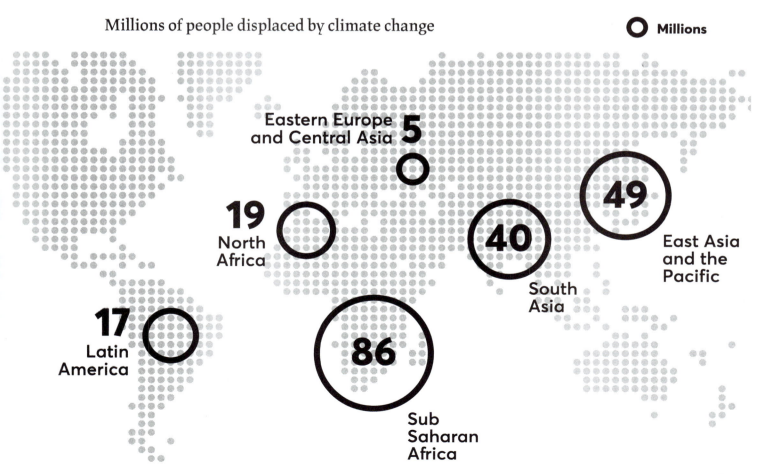

Eastern Europe and Central Asia **5**

19 North Africa

49 East Asia and the Pacific

40 South Asia

17 Latin America

86 Sub Saharan Africa

Rivers carry 10 trillion gallons of water each year, and transport 20 billion tons of sediment to the ocean annually.

2020 Covid Lockdown and Climate

Lockdown and climate

In 2020 the COVID-19 virus spread across the world, and countries went into lockdown. Human activity across the globe was brought to a standstill, and data provided insight on the environmental impact of the reduced activity.

Air quality

Cities documented the best air quality in recent decades. Blue skies were visible in countries such as India and China, which normally experience poor air quality caused by industrial and production activity.

2020 Covid lockdown and climate

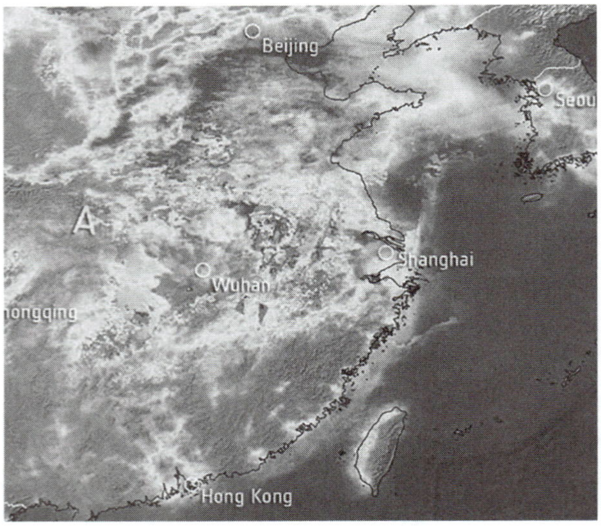

NO_2 emissions late December 2019

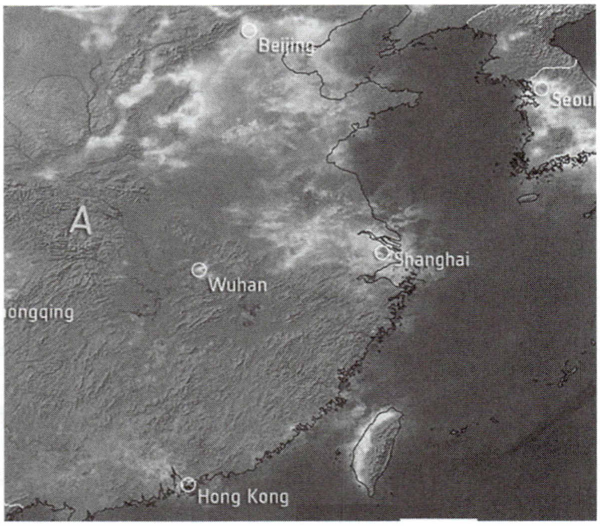

NO_2 emissions early February 2020

Data from 2020 shows:

- 12 percent reduction in NO_2 (Nitrogen Dioxide) across China.
- 31.5 percent reduction in NO_2 across 6 major cities in India.
- 1°C/1.8°F drop in land surface temperature during the day.
- 2°C/3.6°F decline at night.
- 40 percent reduction of harmful airborne particles caused by traffic and energy production across Southeast Asia.

China, Europe, and North America saw emission reductions and improved air quality during the pandemic's first year, while countries such as Sweden saw less dramatic improvements because preexisting air quality contained comparatively lower microparticle levels (PM2.5) of harmful sulfur dioxide (SO_2), nitrogen oxides (NO_x), carbon monoxide (CO) and ozone (O_3). With fewer vehicles on the roads and many industrial activities shut down, emissions that would normally make their way into the environment were absent.

Cities across the world, including New York, San Francisco, Milan, Venice, and Barcelona recorded measured improvement in air quality.

River water

Lockdown forced large-scale industry closures across India from March to September 2020. During that time, river water quality and quantity notably improved:

- 12 percent improvement in the water quality of the Ganges River (which includes sewage waste).
- 50 percent less waste pollution inflow in creeks and rivers in the Mumbai region.
- The Krishna, Cauvery, and Karnataka rivers in central and southern India regained decades-old water quality status.
- All measured rivers across China improved according to their water quality index.
- Venice's murky canal water ran clear, with swimming fish spotted for the first time in years.
- The boat-dominated central section of the River Thames was replaced by seagulls, cormorants, and seals.

Oceans and seas

Oceans and seas cover more than two-thirds of the Earth's surface. During lockdown, fewer boats were launched. The cruise industry came to a total halt, which kept 970,000 liters of graywater from sinks, showers, galleys, laundry, and more, and 110,000 liters of sewage (black water) from being deposited daily into the sea.

The Italian Coast Guard noted underwater ecosystems rejuvenated themselves. Improved water quality also allowed eels, fish, and coral to flourish and increased the presence of whales and dolphins.

In 2020, Thailand, Philippines, Brazil, Florida, the Galapagos Islands, and India saw large turtle spawnings. The absence of human activity allowed them to find deserted beaches to nest in.

Food habits

Food shopping habits, recycling, and waste also changed during the lockdown. WRAP, a UK-based charity, found:

- 63 percent shopped for groceries less often
- 41 percent conducted pre-shopping planning like making a list & checking the fridge and cupboards
- 40 percent cooked more creatively
- 35 percent started checking 'best by date' and chose what to cook accordingly
- 30 percent ensured that leftovers were consumed fully
- 7 in 10 maintained one or more of these behaviors after lockdown

Self-reported food waste

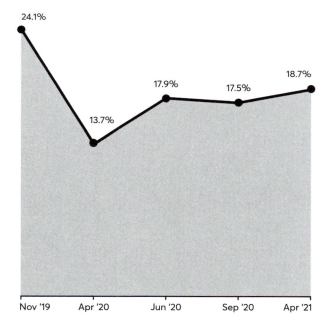

Self-reported food waste dropped almost 50 percent during the lockdown. It has remained lower than the pre-pandemic level.

⊕ **586**

Food Production and Availability

Rising temperatures can push food crops to the limits of survival.

CO_2 concentration, changing rainfall patterns, and extreme weather events affect plant growth. These are already changing the pollination, yield, and nutritional value of crop plants.

Warmer temperatures can cause plants to bloom earlier than usual, which puts a plant and its pollinator out of sync with each other, and then impacts pollination and whether the fruit set. In the US, this could negatively affect over 100 significant crops that rely on pollinators and a larger number of crops in countries worldwide.

When it is too hot or cold for a plant to thrive, it reaches its failure temperature. In the US, the Department of Agriculture lists the failure temperatures for important food crops like beans, wheat, rice, sorghum, corn, and soybeans. Corn, wheat, rice, and soybeans are responsible for three-fourths of the global consumption of calories. The US produces nearly 1/3 of the world's corn and soybeans.

Here are the failure temperatures of these global food crops:
- Beans 32°C/90°F
- Wheat 34°C/93°F
- Rice, Sorghum & Corn 35°C/95°F
- Soybeans 39°C/102°F

Increased concentrations of CO_2 can also alter plant growth and affect a plant's nutritional content. Researchers studying food crops observed that increased concentrations of CO_2 in the air led to a reduction in zinc and iron in some grains and legume crops, reducing the nutritional value of these crops.

⊕ **598**

" Gross National Product counts air pollution and cigarette advertising, and ambulances to clear our highways of carnage. It counts special locks for our doors and the jails for the people who break them. It counts the destruction of the redwood and the loss of our natural wonder in chaotic sprawl.

It counts napalm and counts nuclear warheads and armored cars for the police to fight the riots in our cities. It counts Whitman's rifle and Speck's knife, and the television programs which glorify violence in order to sell toys to our children.

Yet the gross national product does not allow for the health of our children, the quality of their education or the joy of their play. It does not include the beauty of our poetry or the strength of our marriages, the intelligence of our public debate or the integrity of our public officials.

It measures neither our wit nor our courage, neither our wisdom nor our learning, neither our compassion nor our devotion to our country, it measures everything in short, except that which makes life worthwhile.

— Robert F. Kennedy

Agricultural Pests and Diseases

Farms worldwide currently lose between 10 and 15 percent of their crops to pests and diseases.

Climate and agricultural pests

Insect physiology is sensitive to changes in temperature. An increase of 10°C/18°F roughly doubles their metabolic rate. A spike in temperature accelerates insect food consumption, development, and movement.

A new study in Science demonstrates how a 2°C/3.6°F increase in temperature can cause tremendous crop losses from insects. In that scenario Europe and North America face significant potential losses of wheat and maize, while Western Europe stands to lose nearly 75 percent of its wheat crop to pests.

Temperature increases change pest populations, resulting in:

· increased frequency of generations
· increased geographic range
· plant diseases spread by insects
· more likely to survive the winter
· desynchronization of insects and their predators
· plants developing out of sync with insects

Climate and agricultural disease

Fungi affect food crops and generally thrive when the temperature ranges from 20-30°C/68-86°F. As climate change increases global temperatures, a change in fungal diseases is expected along equatorial regions as well.

The Irish Potato Famine was caused by fungal disease—referred to as blight—infecting local crops. There is a strong likelihood that in regions distant from the equator, these diseases can reemerge and impact regional food security.

⊕ 596

Food Insecurity

More than two billion people on the planet are threatened by food insecurity or experience a lack of safe and nutritious food. Higher levels of carbon dioxide in the atmosphere cause rising temperatures, flooding, and degraded land and soil. As a result, the nutritional value and quantity of crop yields—as well as livestock productivity—decrease.

The Center for Strategic & International Studies (CSIS) reports that each 1 degree Celsius rise in mean temperature correlates to a 10 percent drop in crop yields. A heatwave can result in complete crop failure. Poor land management, deforestation, and overgrazing of livestock add to weather-related effects and increase the overall threat to food systems.

Food scarcity will continue to rise, leading to more hunger and malnutrition. Mass migrations away from lands with depleted or unworkable soil for crops and livestock will also increase.

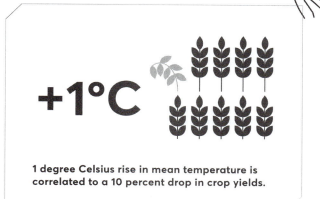

+1°C

1 degree Celsius rise in mean temperature is correlated to a 10 percent drop in crop yields.

The stability of the world's food supply is expected to decline as weather events rise in frequency and intensity. Although this condition is expected to remain highest in low-income countries (58 percent) and lowest in high-income countries (11 percent), food insecurity will continue to exist in every country in the world.

⊕ 067

Land and Soil Degradation

Land and soil degradation is the loss of physical, chemical, or biological qualities that support life. Compared to its state before the Industrial Revolution and factory farming, today more than 75 percent of all the land on Earth is depleted. By 2050, scientists expect that number could reach 90 percent.

Around the world each year, an area equivalent to half of the European Union (4.18 million km²) becomes less productive and resilient. Africa and Asia are the global regions hit hardest by this.

Land degrades through erosion as rock and soil break down and are carried away by wind and rain. This process happens naturally, but extreme weather events make it worse.

As sea levels rise in coastal areas, the neighboring land is lost. What remains can become unusable due to increased salt and other contaminants.

Degradation also occurs through:
· agricultural activities
· animal grazing

· deforestation
· increasing urbanization

Today, 3.2 billion people have experienced the effects of land degradation in some capacity. This has led to a decrease of available food supplies and is often accompanied by an increase in migration.

⊕ 069

Soil Loss

The brown dirt under our feet holds at least a quarter of all global biodiversity and is essential for providing clean water. A teaspoon of soil contains billions of microorganisms. It's estimated that the Earth's soil holds three times more carbon than the atmosphere does.

The importance of soil

95 percent of global food supplies rely on soil, which makes it the difference between life and death for most living beings. Warming the climate by 2°C/3.6°F will push soil to leak more than 230 billion tons of CO_2, which could abruptly tip the planet into irreversible climate change.

The problem

Every minute, roughly 30 soccer fields of soil are eroding or degrading due to:
- agricultural chemicals
- deforestation
- overgrazing

🌐 **579**

PESTICIDE USE

After World War II, large chemical companies focused on the food industry as a market for expansion. Over the following 50 years in the US, insecticide use increased ten-fold while crop losses almost doubled. Pesticide use poisons organisms that create healthy soil on hundreds of millions of acres around the world. For example, earthworms in soil sprayed with pesticides grow to only half their normal weight and fail to reproduce as efficiently as earthworms untouched by pesticides.

WIND POWER

An onshore wind turbine needs to produce electricity for six months in order to offset the energy used to construct it, but after that, for the next 24 years of its lifespan, it produces 100 percent carbon-free electricity.

MASSIVE SOLAR

The Bhadla Solar Park in India is the largest solar farm in the world. It creates 2245 MW of electricity, more than many coal or nuclear plants are capable of. It's located in the desert and the panels are kept clean by robots that operate without using water.

Reductions in Major Crop Yields

According to the United Nations Food and Agriculture Organization (FAO), as many as 811 million people suffered from hunger in 2020. That number is approximately 10 percent of the human population.

As global average temperatures increase, the frequency of droughts and floods is likely to reduce food supplies, while more intense natural disasters and invigorated pests and diseases could further decrease crop yields. There is a wide range of forecasts of the effects of climate change on food yield. Maize, the most important global crop, is forecast to decrease by up to 24 percent. Wheat, the second most important crop, could decrease by 14 percent at 1.5°C/2.7°F of increased warming and as much as 37 percent by 2°C/3.6°F. Soybean harvests could fall between 10-12 percent at 2°C/3.6°F.

At present, the world is able to cope with location-specific droughts or crop failures by sourcing from unaffected areas. The four largest corn exporters, comprising 87 percent of exports, are the US, Brazil, Argentina, and Ukraine. Historically, they have crop failures out of sync from each other because they're geographically distant. But now, all of these areas are expected to see drops in yield between 8 and 18 percent at 2°C/3.6°F and between 19 and 47 percent at 4°C/7.2°F. At 2°C/3.6°F, the risk of all four crop areas failing at the same time is seven percent. If the temperature rises to 4°C/7.2°F, this risk soars to 86 percent.

⊕ **600**

Our people practiced for abundance rather than "sustainability." To me, sustainability means keeping our natural resources on a lifeline until they're eventually gone or until industry has finally had enough and moved on. Practicing for abundance is making sure that your grandchildren won't have to work as hard as you did. It's ensuring that when we leave this garden for them, they will have everything they need.

— Joe Martin

The carbon footprint of one cheeseburger is equal to the carbon footprint of 9 falafel + pita sandwiches or 6 servings of fish and chips.

Food Price Spikes

Food prices depend on changes in supply and demand. While demand is generally stable, supply can vary. Droughts and floods reduce crop productivity and declining farm production threatens food availability, leading to higher prices. So do changes in the cost of marketing and packaging.

```
            PRICE SPIKES
             PLANET EARTH

             RECEIPT: 6631
            DATE: 30062022
        CASHIER: GRETTA JAMES

------------------------------------

  ITEM              GROWTH      PRICE
  MAIZE              ↓12%      ↑90%
  RICE               ↓23%      ↑89%
  WHEAT              ↓13%      ↑75%
  OTHER CROPS        ↓08%      ↑83%


------------------------------------

  SUBTOTAL
  SIMPLE TAX


  TOTAL
------------------------------------

  CASH
  CLIMATE CHANGE       ** IT'S NOT **
                       ** TOO LATE **

====================================

           THANK YOU
        LET'S SAVE THE WORLD
```

By 2030, the growth rates of nine out of 10 major crops will stagnate or begin to decline. At least in part as a result of climate change, their average prices will see a significant increase.

Trade is also a major factor; the UK imports about 40 percent of its food (bananas, tea, coffee, butter, lamb, and more). Most nations' food supply depends upon trade too. In the US, food comes from Canada, Mexico, and other countries. The cost of oil and containers for shipping among countries also determines part of the cost of food.

Food price spikes are amplified by climate change. Average food prices in 2021 are the highest in almost 50 years; for example, the price of coffee increased by 30 percent because of drought, floods, and frost in Brazil. As a result, consumers saw coffee prices rising.

Consumers have already experienced rising prices for bread and pasta because of the reduced production of durum wheat due to droughts in Russia, the US, and Canada—the biggest suppliers of durum wheat. Prices for fruit and vegetables, especially tomatoes, are also rising due to climate change-related problems in Florida and California.

The world has witnessed several food price spikes. In 1973, food prices increased because of the global oil crisis and droughts. In 2008, there was food price inflation related to higher oil prices and droughts in Australia and the US, where policies leading to growing corn for fuel instead of for food drove up animal feed prices. In 2021, food price spikes were similar to those in 1973, but this time extreme weather played a more significant role.

Increasing food prices impact people regardless of income, but in different ways. In low-income households, food prices directly threaten the food supply leading to starvation. In higher-income families, price changes can lead to less healthy diets and a rise in obesity.

⊕ **599**

The Economics of Rising Temperatures

The world stands to lose more than 10 percent of total economic value by mid-century if climate change stays on the currently-anticipated trajectory. If the Paris Agreement and 2050 net-zero emissions targets are not met, widespread and significant economic impacts are likely.

The results of the impacts are likely to include:
- losses to human health and productivity
- damage to infrastructure and property
- consequences to agriculture, forestry, fisheries, and tourism
- increased demand for energy along with less reliable power generation
- stressed water supplies
- disruptions in trade and supply chains

Over the past 20 years, an estimated 500,000 people have died and $3.5 trillion was lost as a result of extreme weather events. In 2017, a poll of economists estimated that future damage from climate change ranged "from 2 percent to 10 percent or more of global GDP per year."

On the other hand, climate change also presents business opportunities and potential positive economic impacts. The Carbon Disclosure Project reported that **225 of the world's 500 biggest companies believe climate change could generate over $2.1 trillion in new business prospects.**

Potential positive economic impacts include:
- Solutions in renewable energy, resilient and green buildings, and energy efficiency
- Hybrid and electric vehicle production, including electric public transit
- Construction of green infrastructure
- Resilient coastal infrastructure
- Carbon capture and sequestration and uses of captured CO_2
- Increased plant-based foods and agriculture
- As Arctic sea ice melts, new shipping lines will open up for trade, and offer more prospects for oil and gas drilling
- Diseases like malaria and dengue will lead to more demand for pharmaceuticals
- Conflicts around the world will lead to more revenue for private security services and military contractors
- New crops resistant to higher temperatures will be developed by biotech companies
- Extreme weather will require more satellites and radar technology

⊕ 604

Economic cost of rising temperature (decreases in GDP)

REGION	+0–2.0 °C *Paris target*	+2.0 °C *Likely range of gains*	+2.6 °C	+3.2 °C *Severe*
North America	-3.1%	-6.9%	-7.4%	-9.5%
South America	-4.1%	-10.8%	-13.0%	-17.0%
Europe	-2.8%	-7.7%	-8.0%	-10.5%
Middle East & Africa	-4.7%	-14.0%	-21.5%	-27.6%
Asia	-5.5%	-14.9%	-20.4%	-26.5%
Oceania	-4.3%	-11.2%	-12.3%	-16.3%

Effects of CO$_2$ on Crop Nutrition

Every plant is made up of a unique mixture of chemicals and nutrients, which is known as the *ionome*. These nutrients come from the soil, but they're dependent on atmospheric CO$_2$ levels to produce carbon-rich sugars and other carbohydrates.

An increase in atmospheric CO$_2$ produces more carbohydrates and fructose in plants, yet simultaneously reduces the nutrients included in the ionome. This can cause a significant drop in plant micronutrient content, including zinc and iron.

60 percent of crop species produce cyanogenic glycoside molecules, which help fend off insect attacks.

A higher concentration of CO$_2$ produces more of these molecules, which can be broken down into cyanide. One important crop is cassava, whose levels of cyanogenic glycosides is already considered high.

The greatest challenge will be faced by food staples like rice and wheat. These crops feed over 2 billion people worldwide. The risk of the rise in the carbohydrate content within these crops is that other nutrients like vitamins will decline. Research conducted in China, Japan, and Australia shows that there is a considerable decline in protein, iron, and zinc in rice grown in a CO$_2$ rich environment.

⊕ **569**

Flooding

There has been a significant increase in the number of floods around the world. Countries with significant flooding events since 1998 include Angola, Australia, Brazil, Belgium, Benin, Canada, China, Congo, Germany, India, Indonesia, Italy, Mozambique, Namibia, New Zealand, Philippines, Rwanda, Turkey, and the US.

In many US coastal areas, high-tide flooding is now 3 to 9 times more frequent than it was 50 years ago. Global sea level is not the only cause of flooding, but it is a contributing factor to some of the worst events.

Three types of floods
- **River Flooding**: When a river overflows and breaches its banks due to excessive rain or unexpected snowmelt upstream.
- **Coastal Flooding**: Extreme weather events draw water inland, where the storm surge can cause coastal flooding and bring ocean or seawater onto the soil.
- **Flash Floods**: In both urban and non-urban settings, heavy rains in a very short period of time can cause unexpected floods.

Why floods happen
Higher temperatures lead more moisture to be absorbed into the air. This results in more flooding due to:
- increased total precipitation in some areas
- heavier precipitation over short time periods
- faster snow melt
- larger storm systems reaching land

⊕ **566**

Sea level increase, 1880-2020

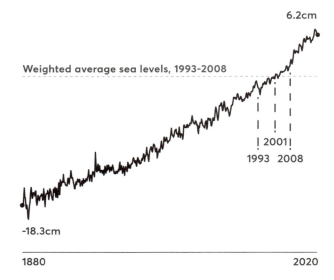

6.2cm

Weighted average sea levels, 1993-2008

| 2001 |
1993 2008

-18.3cm

1880   2020

Contaminated Water and Debris from Flooding

Humans have been writing about large floods since Noah and the ark, and we're used to seeing their impacts on the TV news. But floodwater can carry unexpected debris and contaminants, including:

- human and animal waste
- household industrial waste and chemicals, such as pesticides
- medical waste
- carcinogenic compounds like chromium, mercury, and arsenic
- damaged power lines
- sharp objects including metal or glass
- small snakes and rodents hidden in murky floodwater

Debris in floodwater

Flood debris may include furniture, building materials, cars, trees, and stones, which can overwhelm local waste plants and become a breeding ground for rodents and microorganisms to multiply.

For example, Eastern Belgium's floods in July 2021 accumulated 90,000 tons of piled debris that led to eight kilometers of closed motorway.

Floods and human health

Floodwater can impact human health in many ways:
- skin conditions like infections or tetanus
- gastrointestinal illness including cholera, *E. coli*, or *Salmonella* infection, and other water-borne illnesses like hepatitis
- mosquito-borne diseases such as malaria
- injuries from the removal of debris
- inhalation of dust and mold during clean-up

Impacts on wildlife and livestock

Floodwater impacts wildlife and livestock by:
- destroying farms and ranches
- killing animals as their habitats are submerged
- exposing land- and marine-based plants and animals to contaminated floodwaters
- reducing biodiversity of the affected region
- damaging ecosystems that may not recover
- destroying fish nests and associated displacement

Environmental effects

Turbidity and sediments in floodwaters can damage water quality by leading to the overgrowth of algae. Flooding may also contribute indirectly to pollution by forcing people to buy new products to replace damaged ones. This contributes to increased metal, plastic, and other material waste with its own carbon footprint.

🌐 **588**

ADVANTAGES OF FLOODS (IN UNINHABITED AREAS)

In environments unmodified by humans, floods are a part of the natural course of events and offer benefits by:
- providing nutrients to land and increasing soil fertility by the deposition of silt
- refilling groundwater resources
- moving nutrients into aquatic habitats from the soil
- contributing to the dispersal of species

SUNNY DAY FLOODING

Cities along the US Southeast Atlantic and Gulf coasts are already seeing "sunny day flooding." When tides get about 2 feet above average, streets and storm drains can flood without a cloud in the sky.

Maricopa County in Arizona uses 80,000,000 gallons a day to water its golf courses.

Water Stress

Currently 2.3 billion people or more live in countries suffering from water stress. Water stress is an imbalance caused by a reduction in the supply of fresh water combined with an increase in demand. By 2050, more than half of the world's population (about five billion), is projected to experience water stress.

A region's level of water stress is a reflection of its renewable natural water resources. Water stress is calculated by assessing environmental flow requirements: public water supply, irrigation, and industrial needs. All are necessary to maintain a healthy ecosystem.

Increased water demand

Twelve out of the 17 most water-stressed countries in the world are in the Middle East and North Africa (MENA), which is home to one-quarter of the world's population. These countries withdraw more than 80 percent of their water supply on average every year.

Growing populations and rising temperatures continue to increase the demand for water across the globe. Manufacturing is also expected to double the demand for water resources between 2000 and 2050.

Reduced water supply

Reduced water supply stems from:
- **Droughts:** Warmer air increases evaporation across water and soil, creating more frequent and severe droughts. Since 2000, droughts have increased by 29 percent.
- **Reduced snowpacks:** Higher temperatures mean less snow and more rain. This reduces snowpacks that add fresh water to rivers and streams and replenish drinking water supplies. Since 1915,

snowpack in the western US has declined by 21 percent. Estimates project the Sierra Nevada mountain range in the US could lose between 30 and 64 percent of its snowpack by 2100.

Contaminated water

Globally, one in three people do not have access to safe drinking water. Contaminated water is caused by various conditions:
- Heavy rainfall creates surface runoff that carries contaminants into lakes and streams. Contaminants are toxic to people, fish, and wildlife.
- Fertilizer runoff encourages algal blooms that may diminish the oxygen levels in water, increasing treatment costs and making water untreatable.
- Warmer water holds less dissolvable oxygen, lessens viability of aquatic systems, and reduces water availability and fish stocks.
- Rising sea levels contaminate freshwater-bearing rocks (aquifers) with saltwater, requiring costly and energy-intensive desalination processes to make water usable.

Precipitation variability

Changing ocean temperatures alter atmospheric circulation patterns, weather patterns, and the location of precipitation. The increasing unpredictability of rainfall combined with greater intensity may create flooding conditions, and increase the possibility of runoff water contamination.

⊕ **587**

Water-stressed countries

Stress index normalized for comparing countries, based on scaling up sub-catchment risk estimates

RANK	NAME	STRESS INDEX	RANK	NAME	STRESS INDEX
1	Bahrain	5.00	17	Macedonia	4.70
1	Kuwait	5.00	18	Azerbaijan	4.69
1	Qatar	5.00	19	Morocco	4.68
1	San Marino	5.00	20	Kazakhstan	4.66
1	Singapore	5.00	21	Iraq	4.66
1	United Arab Emirates	5.00	22	Armenia	4.60
1	Palestine	5.00	23	Pakistan	4.48
8	Israel	5.00	24	Chile	4.45
9	Saudi Arabia	4.99	25	Syria	4.44
10	Oman	4.97	26	Turkmenistan	4.30
11	Lebanon	4.97	27	Turkey	4.27
12	Kyrgyzstan	4.93	28	Greece	4.23
13	Iran	4.91	29	Uzbekistan	4.19
14	Jordan	4.86	30	Algeria	4.17
15	Libya	4.77	31	Afghanistan	4.12
16	Yemen	4.74	32	Spain	4.07

We're fighting for soil, land, food, trees, water, birds. We're fighting for life.

— Gregorio Mirabal

Dust Storms

The world is getting hotter and drier. When combined with the effects of desertification, increased wind events, and heat waves, dust storms are becoming a more common phenomenon on the planet.

Also known as *siroccos* or *haboobs*, dust storms can be unpredictable and highly destructive. A moving wall of sand, dust, and debris picked up from the parched ground, a dust storm can be miles long and several thousand feet high.

Dust storms are historically most common in the Middle East and North Africa. Now they are increasingly common in the US, where they have doubled in frequency since the 1990s.

The rise in dust storms around the world is increasingly being linked to climate change. The US National Oceanic and Atmospheric Administration (NOAA) has linked the rise in ocean temperatures caused by climate change to the increase of dust storms in the southwestern US. Rising ocean temperatures in the Pacific Ocean produce more winds that blow across the region and dry out soils at an increased rate.

Following a storm, dust can remain in the air for hours or even days, and may then be carried to other areas. The resulting dust also leads to negative human health impacts, including Valley Fever created by soil-born fungi and increased Intensive Care Unit (ICU) hospital visits due to respiratory problems.

77 percent of the world's countries are directly affected by dust storms. 23 percent are considered dust storm source areas, meaning that they are dry enough to cause new sandstorms.

⊕ **075**

> *We always overestimate the change that will occur in the next two years and underestimate the change that will occur in the next ten. Don't let yourself be lulled into inaction.*
>
> — Bill Gates

THIRSTY LAWNS

On average, American households use 1,450 liters of water **per day**. In arid regions like the southwestern US, as much as 60 percent of that water is used for watering lawns. Water usage in the US overall amounts to 34 billion liters per day—the equivalent of 13,600 Olympic-sized swimming pools with each holding 2.5 million liters of water.

Hot Droughts

Droughts and heat waves can amplify each other. A heat wave can cause a cascading effect, leading to crippling drought. And often it can happen in the other direction as well.

Droughts and heat waves combined can lead to:
- water scarcity for plants, humans, and other animals
- substantial agricultural loss and an associated increase in food insecurity
- increased air pollution
- an increase in the frequency, intensity, and scope of wildfires

A drought occurs when there are shortfalls in precipitation over extended periods of time. The impact of a drought is heightened by heat waves, which are periods of several days to weeks of hot weather exceeding the norm for the area. Combined, these "hot droughts" do more damage than each of the two weather extremes do when they occur on their own.

Droughts remove soil moisture, leading to less evaporation. This multiplies the effect of a heat wave because surface cooling is reduced. Heat waves create more demand for water, amplifying the impact of a drought as well.

As the climate changes, hot droughts are more common. A drought today is 4°C/7.2°F warmer than a similar drought 170 years ago.

Hot droughts have occurred worldwide, including in the Southwestern United States and parts of Canada, Europe, and northern Africa.

⊕ **076**

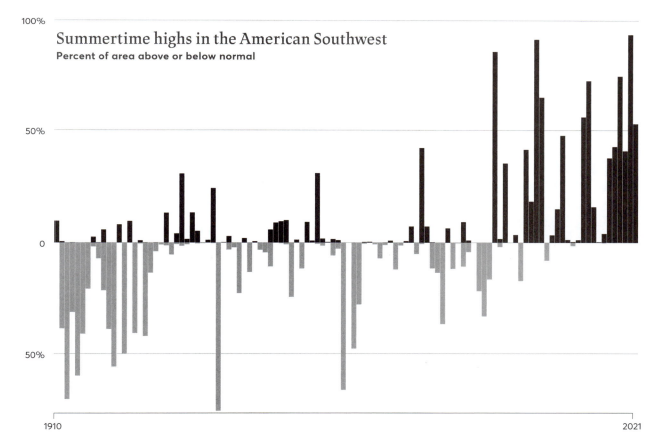

Summertime highs in the American Southwest
Percent of area above or below normal

100%

50%

0

50%

1910   2021

Desertification

Desertification happens when fertile land becomes desert. The land can't sustain the plant life it had previously and the change is considered irreversible.

Drier soil increases the intensity of heat waves, which further dries out the soil. More carbon dioxide is released into the atmosphere as the land is covered by less and less vegetation, leading to rising global temperatures.

> **There's often a cascade effect, in which historically fertile land begins a one-way path to becoming a desert.**

Desertification also leads to a loss of biodiversity. Wildlife and plant life can't quickly adapt to such changes in their ecosystem, leaving room for invasive species to move in.

The dryness of the soil also means an increase in sand and dust storms. Their impact can be devastating to human health, open water sources, and transportation and energy infrastructure. Dust storms further add to an area's aridity by making rainfall even less likely.

The combination of drought and heat waves increases the risk of wildfires. The drier and warmer conditions give fires a chance to escalate faster and burn more intensely. Degraded soil contributes to the spread of the fires because the land doesn't retain the same levels of water and nutrients.

Desertification is occurring around the world and affecting half a billion people. Today the effects are felt in the American Southwest, North Africa, Northern China, Russia, and the northeastern corner of Brazil in South America. In northeastern Brazil, desertification is endangering the 53 million people living in what is already one of the most populated dryland areas in the world. By 2050, the number of people living in drylands (arid, semi-arid, and dry sub-humid areas) across the planet is expected to increase by 43 percent to 4 billion. By 2100, drylands are expected to cover over 50 percent of the Earth's land surface. Parts of Australia, North Africa, the western United States, and northeastern South America may become uninhabitable by humans by 2100.

🌐 **066**

> *There is true relief in the fellowship of others who understand and who are grieving.*
>
> — Dr. Margaret Klein Salamon

Loss of Wetlands and Marshes

Coastal wetlands provide essential natural infrastructure for the 2.4 billion people worldwide who live near the coast. As sea levels rise, the ocean is encroaching on these wetlands.

What are wetlands?

Wetlands include some of the most carbon-dense eco-systems on Earth, such as salt marshes, seagrass beds, and mangroves. The National Oceanic and Atmospheric Administration (NOAA) classifies wetlands into five general types:

- marine (ocean)
- estuarine (estuary)
- riverine (river)
- lacustrine (lake)
- palustrine (marsh)

Why do wetlands matter?

- Wetlands provide almost all of the world's fresh water for consumption.
- Wetlands filter impurities from runoff before it flows into the ocean.
- Wetlands act as a natural sponge and absorb floodwaters, protecting people, property, infra-structure, and agriculture from damage.
- Peatlands store twice as much carbon as forests.
- Wetlands are home to more than a third of the United States' threatened and endangered species, with many other species depending on wetlands for survival.

Wetlands and marshes are "drowning"

Rising sea levels are submerging coastal wetlands. Approximately 64 percent of the world's wetlands have been lost since 1900, with 35 percent lost from 1970 to 2015. Southern California has already lost three-quarters of its salt marshes, and salt marshes across California and Oregon could disappear entirely by 2100.

What will happen if wetlands disappear?

- Coastal regions will experience greater damage from hurricanes and typhoons
- Increased frequency of high-tide or "sunny-day" flooding, which occurs when high tides flood low-lying city streets
- A decline in freshwater quality and supply
- The release of carbon will contribute to increased greenhouse gases in the atmosphere
- Extinction of animals that depend on wetlands

Between 13 and 18 percent of worldwide marshlands are on the Ramsar List of Wetlands of International Importance, which are protected sites.

🌐 **080**

In 2021, electric vehicles outsold diesel ones in Europe for the first time.

Extreme Precipitation

Intense downpours are increasing around the globe, and the number of record rainfall days is growing. Storms with over 50mm of precipitation per day have increased 20 percent in the United States over the last century. Similar trends have been found all over the globe.

Intense rainfall outcomes

Engineers design lakes, drainage systems, and streets to handle the maximum rain swell typical of that region. When heavier-than-normal rainfall occurs, it can over-whelm the area's infrastructure.

When excessive precipitation occurs in an area built over an ancient lake bed that's been dry for centuries, the development and infrastructure can quickly become overwhelmed and submerged.

Relentless rainfall events are becoming the norm:

- In December 2015 Chennai, India received 494 mm of rain in a twenty-four-hour period—more rain in one day than in a typical rainy *month*.
- In July 2021, extreme rainfall hit the Henan Province in China, causing flash floods, 302 deaths, and $17.7 billion in losses. In five days, total rainfall was 720mm—more than the city's *annual* average.
- Western Germany and eastern Belgium also experienced extreme downpours in July 2021 on ground already so saturated it caused flooding, landslides, and over 200 deaths.

⊕ **574**

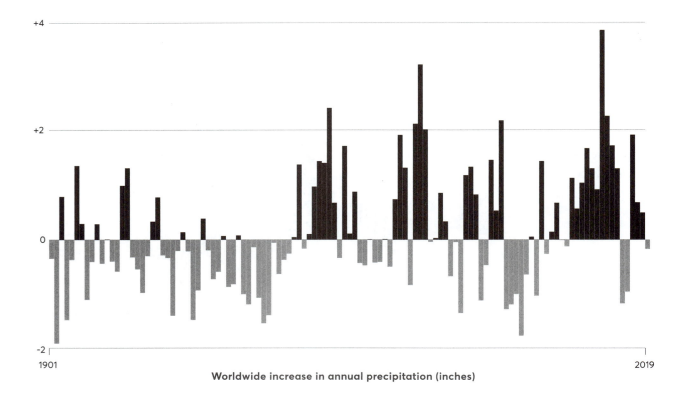

Worldwide increase in annual precipitation (inches)

Wildfires

> *My sense of place went up in a puff of smoke on June 30th. My grandmother told me that her grandmother told her that we'd lived here forever. I've seen changes in ecosystems, I've seen less water, I've seen the trees struggling with drought and heat anxiety. I can't even tell the weather anymore.*

— **Patrick Michell, chief of Kanaka Bar Indian Band and resident of Lytton, Canada,** which burned down in one day during the Pacific Northwest's record-breaking heatwave in 2021.

Infernos like the one that leveled Chief Michell's home in British Columbia are becoming more common. The rise in temperatures worldwide intensifies the arid, hot, and windy conditions that spark uncontrollable bush fires or wildfires, which accelerates climate change. Burning some of the planet's most effective carbon sinks—forests, meadows, peatlands, and grasslands—transforms them into massive carbon emitters.

In 2020, California wildfires emitted more than 91 megatons of CO_2, about 30 megatons more than the state's annual average emissions from power production. Over that same time period, more than one gigaton of emissions was released during fires spanning Australia, the Amazon, and Siberia.

As this cycle builds momentum, it can generate even more destructive wildfires known as "megafires," which can:

- spawn flames 200 feet high
- produce their own weather systems in the form of fire clouds, dry lightning, and fire tornadoes
- burn down entire towns
- ruin local economies
- reduce local biodiversity of flora and fauna
- pollute the air for hundreds or thousands of miles in any given direction

Unless worldwide emissions are cut significantly, "high fire danger" days when conditions are ripe for wildfires could increase globally by 35 percent by mid-century. The most vulnerable areas will include southern Africa, southeastern Australia, the Mediterranean, and the western US.

⊕ **583**

50%
Wildfires affect a yearly estimate of 400 million hectares of land worldwide—about half the size of the US.

What Is Biodiversity?

Tiny creatures might be invisible, but they matter.

About 25 percent of all the life on Earth is hosted in the soil. A cupful of soil holds as many microorganisms as there are humans on Earth. And the typical human hosts 38 trillion bacteria in their body.

Biodiversity refers to the variety of all forms of life on Earth—not just the "cuddly" mammals in photos and at the zoo, but all lifeforms, including:

- animals
- plants
- microbes
- viruses
- fungi

Impacts of biodiversity loss

Lifeforms on Earth have evolved over billions of years. Many have grown to be co-dependent with other species.

Declining biodiversity can upset the delicate balance within ecosystems, causing domino effects where the extinction of one species can lead to declining numbers in another. Since 2009, 14 percent of reefs have been lost, leading to a steep decline in marine biodiversity.

Declining biodiversity can upset the delicate balance within ecosystems.

Declining biodiversity can also pose more direct dangers to humanity, including:

- reduced availability of clean water
- food shortages from declining populations of fish
- loss of resources from forests, such as oxygen, plant medicines, and food
- loss of livelihood for communities depending on natural resources and ecotourism
- less diverse and healthy crop stocks due to declines in pollination

🌐 **065**

Biodiversity Loss and Climate Change

Climate change forces species to adapt either through shifting habitat, changing life cycles, or the development of new physical traits. When a species cannot adapt, it perishes. Climate change threatens almost 20 percent of all at risk and endangered species.

Each tiger in this chart represents 60 wild tigers. Only 3200-3600 wild adult tigers remain

Climate change drives biodiversity loss by:

Ocean warming and acidification
Corals are vulnerable to rising temperatures. Ocean acidification can make it harder for shellfish and corals in the upper ocean to form shells and hard skeletons.

Rising global temperatures
These can alter ecosystems over longer periods by changing what species live within them. Evidence indicates that increased evaporation since the 1990s have resulted in 59 percent of vegetated areas showing pronounced browning and reduced growth rates worldwide.

Worsening weather
The increased intensity and frequency of fires, storms, or periods of drought also impacts biodiversity.

In Australia, the bush fires of 2019-2020 were intensified by climate change and destroyed 97,000 km² of forest and surrounding habitat. As a result, the number of species threatened in that area increased by 14 percent compared to before the fires.

Other drivers of climate change also drive biodiversity loss
- Converting land for agriculture has caused 70 percent of global biodiversity loss.

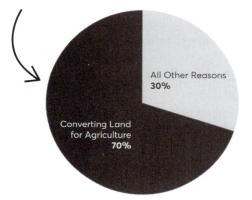

- Monocropping and pesticides reduce the variety of crops and insects (including pollinators) and lead to extinction.
- Deforestation can directly lead to biodiversity loss when animal species that live in the trees no longer have their habitat, cannot relocate, and therefore become extinct.
- Urbanization and road construction cause biodiversity loss mostly from habitat loss and fragmentation. From 450,000 km² in 2000, the amount of urban land near protected areas around the world is expected to increase by more than three times between 2000 and 2030.
- Expanding deserts both drive and are driven by biodiversity loss.

⊕ **074**

Impact on Forests

A single tree hosts as many as 2.3 million organisms...

Covering nearly a third of the planet's surface, forests provide plants, wildlife, and humans with a wide range of benefits, including:

- water and air purification
- climate regulation through carbon storage in the forest's trees and soil
- shelter and protective habitats
- medicinal ingredients
- protection of biodiversity
- livelihoods for people living in or near the forest
- products like lumber and paper
- a place supporting spiritual and mental health
- providing seeds for future generations of trees

A single tree hosts as many as 2.3 million organisms, including microbes, insects, birds, and mammals. Nearly 25 percent of the global population (1.6 billion people) depend on forests for their livelihoods, while 80 percent of the world's terrestrial biodiversity can be found in forests.

Climate change from human and natural sources warms the planet, causing more trees to die. Longer periods of drought from extreme heat waves are more common and more intense. As a result, forests are more susceptible to insects and diseases, leading to more dead trees.

Dead trees added to hot, dry conditions cause forests to become more susceptible to uncontrolled wildfires. When fires burn deep into the soil of the forest floor, the seeds in the soil are often destroyed, leading to prolonged damage to forests.

Burning forests transition from absorbing carbon dioxide to emitting it, adding to the climate change problem. Between 2010 and 2020, more than 162 million trees died in California alone. Climate change-related stress appears to be the main cause.

⊕ **077**

...and the forests sustain nearly 25 percent of the global population (**1.6 billion people**).

Ground-Level Ozone

The Northern Hemisphere has experienced a 5 percent per decade increase in ground-level ozone.

Ozone is a gas composed of three oxygen atoms. It occurs naturally but is also man-made. Ozone is at two levels in the atmosphere:

- **Stratospheric ozone** occurs naturally in the upper atmosphere and is necessary for life. This type of ozone forms a protective layer that shields humans from some of the sun's harmful ultraviolet rays.
- **Tropospheric (ground-level) ozone** is man-made, a harmful air pollutant, and a key ingredient in smog.

Ground-level ozone is the result of air pollution mixing with sunlight. When nitrogen oxides combine with volatile organic compounds, the result is ground-level ozone. This is created by emissions from vehicles, industrial power plants, refineries, and chemical plants.

Ground-level ozone is dangerous to human health at levels over 70 parts per billion. It is often worse in the warmer months, but can still reach high levels in winter. Because ozone is airborne, it can be transported long distances by the wind. Even rural areas are susceptible to high ozone levels as a result.

Breathing ground-level ozone can be harmful to human health. People suffering from asthma and other respiratory diseases are at a greater risk of negative impact.

Elevated exposure to ozone can also negatively impact sensitive plants. Ozone damages plants by entering leaf openings and burning the plant tissue during respiration.

⊕ **570**

Increased Ozone Inhibits Photosynthesis

Photosynthesis is the foundation of life on Earth and the starting point of the food cycle. During photosynthesis, plants release oxygen humans and animals breathe.

But air pollution damages plants and slows down photosynthesis. The pollutants come from industrial facilities, vehicle exhaust, gasoline vapors, and chemical solvents. They take the form of nitrogen oxides (commonly referred to as NO_x) and volatile organic compounds (known as VOCs).

The NO_x and VOCs undergo a chemical reaction with sunlight and create ground-level ozone.

$$NO_x + VOCs + sunlight \longrightarrow ground\text{-}level\ ozone$$

Many plants absorb ground-level ozone. When ozone enters the plants' tissues, it slows the process of photosynthesis and restricts their growth. That makes them more susceptible to damage from diseases, insects, and severe weather.

Less photosynthesis also means the plants convert less carbon dioxide to oxygen for humans to breathe.

The implications of increased air pollution and less photosynthesis are easily imagined: There will be less plant material to consume throughout the animal kingdom, less conversion of CO_2 to oxygen, and a cycle of increased air pollution and greenhouse gases.

As a result, it has been said that ground-level ozone causes more damage to plants than all other air pollutants combined.

⊕ **364**

Don't take our word for it.

Visit **www.thecarbonalmanac.org/364** to check out this article's sources, relevant links, and updates.

Impact to Peatlands

The largest natural storehouses for carbon on land are peatlands, also known as wetlands and swamps. The United Nations Environment Program determined that the world's peatlands store more than twice the amount of carbon dioxide as is stored by the world's forests.

These wetlands cover 3 percent of the Earth's land surface. They absorb and store more than one-third of a gigaton of carbon dioxide per year—more than all other forms of vegetation on Earth combined. Peatlands worldwide hold more than 550 gigatons of carbon (42 percent of all soil carbon) in more than 3 million square kilometers.

Russia, Canada, Indonesia, and Alaska have the largest known tracts of peatland, though they are found in most countries. Not all peatlands on Earth have been mapped. Peatlands develop when constant water-soaked conditions slow down the composition of vegetation. A "peat" or dense soil of dead plant matter forms over thousands of years and can be several meters thick. Peatlands further from the equator are commonly treeless open bogs while peatlands closer to the equator are swamps beneath tropical forests.

Damaged peatlands emit stored carbon at the rate of 5 percent of all global human-made CO_2 emissions. Thawing permafrost in northernmost peatlands has the potential to release up to four times more CO_2 than all of the human-made CO_2 emissions combined. Nearer the equator, peat fires in Indonesia released nearly 16 million tons of CO_2 a day—more than the daily emissions of the entire US economy.

Higher temperatures and drought have the potential to cause even more damage. Drained peatlands release 1.9 gigatons of CO_2 annually.

⊕ **084**

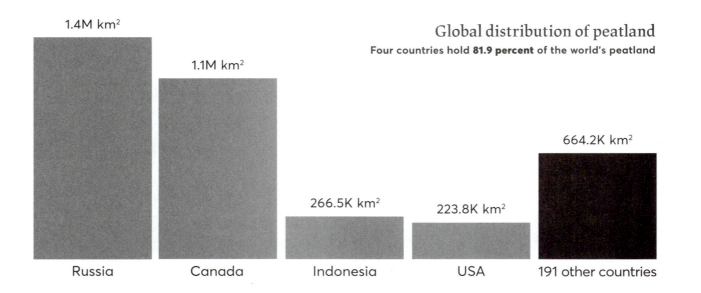

Global distribution of peatland
Four countries hold 81.9 percent of the world's peatland

Russia	Canada	Indonesia	USA	191 other countries
1.4M km²	1.1M km²	266.5K km²	223.8K km²	664.2K km²

This isn't about when. It's about now. It's not about who. It's about us. And it isn't about cost. It's about an existential necessity.

— Patrick Odier

Carbon and the Oceans

The oceans breathe.

We never see it, but the surface of the oceans interact with the atmosphere, absorbing and releasing CO_2.

Of the carbon dioxide that is released by humans, approximately a quarter will be absorbed by the oceans. For millions of years, the oceans acted as a buffer, removing carbon from the air when it was high, and releasing some when it was low. Historic values pre-Industrial Revolution show a mildly positive net flux, indicating that more CO_2 was released from the ocean into the atmosphere—which is not true for today's oceans.

The oceans breathe.

As the ocean and atmosphere are in constant exchange, some of the carbon dioxide pumped into the atmosphere by human activity will continuously be absorbed by the oceans. This process will continue until rising temperatures slow down the oceans' water circulation and its ability to absorb more carbon.

The water in the ocean circulates, with warm water rising, then cooling, then falling. Surface water absorbs some carbon dioxide, and then falls to the depths, replaced by currents of water that have not yet been exposed. But as the atmosphere heats up, this circulation changes.

First, the surface water is less able to absorb CO_2. As atmospheric CO_2 increases, the oceans have traditionally absorbed more of it, but as water temperature increases, its ability to dissolve CO_2 *decreases*.

Second, as the surface temperature increases, it's harder for wind and currents to mix the water—it stratifies. As a result, the water that was rich in carbonates at the bottom tends to stay at the bottom. If the water isn't mixing and circulating from top to bottom, then carbonate deposits will increase.

Water from deeper depths is generally rich in carbonates, which come from limestone or dead marine organisms on the ocean floor. Currents have traditionally brought these carbonates up to the surface, allowing the ocean to absorb more carbon as well as giving marine life the environment it evolved to live in. As temperatures increase further, ocean water will undergo stratification, which will make it harder to absorb carbon.

Once the rate of circulation has declined, the surface water's ability to absorb CO_2 will slow and could stop. When saturation levels peak, more CO_2 will remain in the atmosphere, which in turn will further accelerate a rise in temperatures and add to the warming of the planet.

🌐 **676**

Mining one bitcoin emits 191 metric tons of carbon—about 13 times as much as mining the equivalent value of gold.

Bleaching and Loss of Coral Reefs

Coral reefs are found on less than 1 percent of the ocean floor but more than 25 percent of marine biodiversity depends on the reefs. Coral reef ecosystems are under global threat from climate change.

> **Coral reefs are found on less than 1 percent of the ocean floor but more than 25 percent of marine biodiversity depends on the reefs.**

First emerging on Earth at least 500 million years ago, today coral reefs provide direct support for more than 500 million people around the world. The majority of these people live in developing countries.

Reefs are home to thousands of species of fish, mollusks, crustaceans, and other sea creatures. Fisheries dependent upon reefs represent one small part of what makes reefs valued as high as $2.7 trillion annually.

Human-caused climate change leads to:
· warmer ocean temperatures
· ocean acidification
· increased storm frequency and intensity

All of these impacts of climate change endanger coral reefs.

An estimated one-third of the excess carbon dioxide in the atmosphere is absorbed by the world's oceans, lowering their overall pH levels and reducing calcification. Increasing ocean temperatures cause mass coral bleaching events and spread infectious diseases within the reefs.

When temperatures rise, coral polyps become stressed and expel the algae or *zooxanthellae* living in the polyp tissue. The algae provide both color and food to the coral so when it leaves the coral becomes white or "bleached." Bleached coral is not dead but it is experiencing high levels of stress. It is more vulnerable to disease or breakage from storms, which makes it more likely to die.

Overfishing, marine and land-based pollution, and coastal development have already stressed the world's coral reefs. Adding human-caused climate change that raises the temperature and acidity of ocean waters has put coral reefs around the world in decline. As much as one-third of the reef-building species supporting the marine biodiversity of reefs are in danger of extinction. ⊕ **592**

> *If you protect the coral reef, you protect the ocean ecosystem.*
>
> — Chiahsin Lin

FOSSIL FUELS

According to the Global Fossil Fuel Divestments Commitments Database, approximately 1500 large institutions with $39.2 trillion in assets are publicly divesting from fossil fuels.

Shoreline Erosion

Shorelines constantly change. The coasts are dynamic and some amount of shoreline erosion is natural as oceans and landforms interact.

The 120-meter rise in sea levels since the last ice age has created today's coastlines. The accelerating rate of sea-level rise caused by climate change will have a growing impact on coastlines worldwide. More powerful hurricanes and storm systems will create violent storm surges that submerge and reshape coastal regions. The impact of coastal degradation on biodiversity is already underway.

Sandy beaches occupy 31 percent of the world's ice-free coastline. These beaches are particularly threatened by shoreline retreat, with many of them eroding at a rate of half a meter per year. For 4 percent of beaches erosion exceeds five meters per year and for 2 percent the number exceeds ten meters per year. Overall, half of the world's most vulnerable sandy beaches will be wiped out by climate change by the end of the century. The increased rate of shoreline erosion will also result in more frequent flooding of inland regions.

> **Overall, half of the world's most vulnerable sandy beaches will be wiped out by climate change by the end of the century.**

More frequent storm surges will have an impact on coastal communities and force their relocation inland. While wealthier communities might be able to implement adaptive resilience planning and flood mitigation technology, the most heavily impacted people will be those that depend on coastal regions for their livelihoods. 600 million people live less than 10 meters above sea level, and 40 percent of the world's population live within 100 km of the shoreline.

🌐 **078**

> *The ocean has protected us from the worst effects of climate change by absorbing more than 90 percent of the excess heat already caused by human emissions since the dawn of the Industrial Revolution and by taking up an amount of carbon roughly equal to that emitted by global transportation each year.*

— Peter de Menocal

Search the web, plant a tree

The Carbon Almanac teamed up with Ecosia to make your online searches more powerful. Visit **www.thecarbonalmanac.org/search** to install a simple extension that plants a tree every time you do some web searches. It's free. Just as fast and even easier than Google, but it makes a difference, every day.

143,000,000 trees planted as of 2021.

The Impact of Thawing Permafrost

Frozen land near the Arctic circle is called permafrost. It develops when ground temperatures are below 0°C for two or more years. Approximately 15 percent of the land in the Northern Hemisphere is frozen solid. Permafrost is found in Russia, Canada, Alaska, Iceland, the Himalayas, and Scandinavia.

Permafrost encompasses large amounts of dead and decayed plants and animals preserved in the soil at varying stages of decay. This frozen matter contains nitrogen, carbon, carbon dioxide, and methane.

Permafrost contains about 1,500 gigatons of carbon, which is four times the total amount humans have emitted since the Industrial Revolution.

> **Permafrost contains about 1,500 gigatons of carbon, which is four times the total amount humans have emitted since the Industrial Revolution.**

As global temperatures rise due to climate change, the permafrost thaws. Microbes then become active in the warmer soil and begin devouring the carbon. They emit CO_2 and methane. Bubbles of CO_2 and methane that were frozen in the soil are also released as the soil softens.

Arctic regions are warming two to three times faster than the global average, already heating up 2°C/3.6°F above their pre-industrial temperatures. This rapid increase is predicted to double by 2050.

Wildfires related to extreme heat are increasingly frequent and intense. They amplify the cycle of thawing, greenhouse gas emissions, and Arctic warming as they consume even more carbon-rich land, sending more CO_2 into the atmosphere through combustion.

Earth's entire climate system is impacted by permafrost thaw, as the release of long-stored greenhouse gases intensifies climate change worldwide.

🌐 **486**

FASTER AT THE POLES

Arctic regions are warming two to three times faster than the global average.

The Shrinking of Glaciers

The majority of the world's 220,000 glaciers are shrinking or disappearing. Melting glaciers are responsible for at least 21 percent of global sea-level rise over the last 20 years.

Glaciers cover just 10 percent of Earth's landmass but hold 70 percent of Earth's freshwater. The world's glaciers lost about 267 billion tons of ice every year between 2000 and 2019. Most of this glacial meltwater ends up in the ocean.

Seven glacial regions produce 83 percent of global meltwater. One-quarter of it comes from Alaska, where there is a high concentration of glaciers, rapidly increasing temperatures, and decreasing snowfall.

The ice in glaciers was formed thousands of years ago. Once covering parts of every continent, the remaining glaciers are left over from the end of the last Ice Age. They are now mostly found in Arctic areas and high-altitude mountaintops.

Each year, glaciers gain mass through snowfall that becomes packed down on top of the frozen ice from thousands of previous years. This cold weather process offsets the meltwater loss from the glacier during warmer weather. As snowfall declines and temperatures increase, the balance of glacial mass is lost.

In the western United States, South America, India, and China, the water from glacial runoff in the summer creates an annual water supply for hundreds of millions of people and surrounding ecosystems. With glaciers shrinking and disappearing,

Glacial loss per year (gigatons)

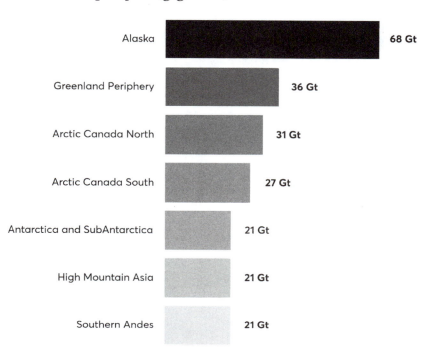

Glaciers gain mass through snowfall that becomes packed down on top of the frozen ice from thousands of previous years.

the people and habitats in these areas are endangered.

For millennia, 90 percent of the solar radiation hitting the bright snow and ice reflected back to space. But as snow and ice melt, the rising ocean and the darker uncovered land now absorb more radiation and then release it as heat into the atmosphere. Temperatures rise as a result. As the cycle shifts, even more ice melts.

When Glacier National Park was created in 1910, it included 150 glaciers. Today, there are fewer than 30 glaciers left, and most of those have shrunk by at least two-thirds. By 2050, the park is projected to lose most if not all of its glaciers. Since 1912, 80 percent of the snows of Mount Kilimanjaro have disappeared. Most of the glaciers in the eastern and central Himalayas are projected to be gone by 2035.

⊕ **593**

Snowfall and Melting Arctic Ice

Arctic sea ice moderates the climate of the planet by keeping polar regions cool. Its bright surface reflects 80 percent of sunlight back into space and acts as a protective barrier. But Arctic sea ice is now declining at a rate of 13 percent per decade.

When that barrier disappears, increased evaporation means additional vapor pumped into the atmosphere in the form of rain, humidity, and snow. More extreme weather is the result.

Researchers found that every square meter of winter sea ice lost in the Arctic resulted in a 70 kg increase in evaporation. This contributed to a specific weather event called "Beast from the East," a historic snowfall in February 2018 that immobilized many parts of Europe and caused record-breaking accumulation in Rome, Italy. Water vapor traveling south from the Arctic carried a unique geochemical imprint that came from the warm, ice-free surface of the Barents Sea between Norway, Russia, and Svalbard.

⊕ **572**

Marine Heat Waves

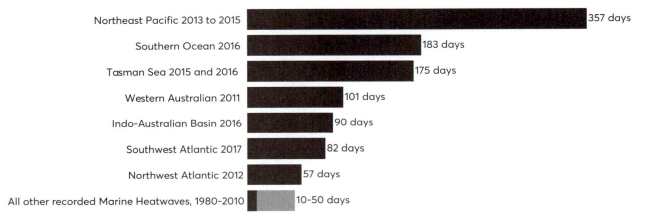

Heat waves on land are easy to discover and difficult to live through. But they can occur over the open ocean as well and can have a similarly dangerous impact.

A marine heat wave (MHW) is an abnormal and often undetected warm water event. In order to qualify as an MHW, the duration must last for five days or longer, record temperatures greater than the 90th percentile of 30-year local measurements for that time of year, and occur in a specific region.

Marine heat waves can force a change in the underwater flora and fauna, which can be harmful to regional biodiversity. Algae blooms often result from MHWs since the warmer conditions are conducive to such growth, but certain algae produce toxins that can have a harmful effect on the developmental, neurological, and reproductive capacities of marine organisms.

Across the globe, oceans have been consistently warming, and over 30,000 distinct MHWs have occurred in the last four decades. While MHWs are a local phenomenon and only last for a few days, the damage they can cause during that short period is tremendous and the recovery can be long and sometimes impossible. Recently, the duration of marine heat waves has been increasing.

⊕ **573**

Hurricanes, Typhoons, and Cyclones

As carbon levels in the atmosphere increase, Earth's temperature increases. This leads to an increase in water evaporation, which leads to more moisture in the atmosphere. For every 1°C/1.8°F of warming, the atmosphere holds an additional 7 percent of moisture. This is known as the Clausius-Clapeyron relation.

As water evaporates from the oceans, heat transfers from water to air. When storms pass over warm oceans, they suck in more water vapor and heat. This leads to stronger winds, more rainfall, and increased flooding. Hurricanes now produce between 4-9 percent more rain than before the Industrial Revolution.

Climate change amplifies hurricanes, and they weaken more slowly when they make landfall. They will cause more damage as they linger over areas longer.

Between 15 and 38 percent more rainfall was attributed to the rise in global temperatures during hurricane Harvey in August 2017. This is at least double the 7 percent Clausius-Clapeyron limit.

If the temperature increases to 3-4°C/5.4-7.2°F, then rainfall from hurricanes is forecast to increase by as much as 33 percent and wind speeds could be as high as 94 km/h faster.

Since 1975, the typhoons that strike East and Southeast Asia have intensified by nearly 15 percent. There have been nearly three times as many category four and five storms.

If global temperatures rise by even just 2-3°C/3.6-5.4°F, storms will get worse. A storm like Cyclone Yasi, which struck Australia in 2011, would drop 35 percent more rain. Cyclone Gafilo, which hit Madagascar in 2004, would have dropped 40 percent more.

🌐 **567**

Global weather-related disasters

Flood
170 events

Extreme weather
85 events

Landslides
22 events

1969   2019

---

1 CUP OF COFFEE

According to UNESCO's Institute for Water Education, about 39 gallons of water are needed to produce one cup of coffee, compared to nine gallons to produce a cup of tea.

Energy Production and Negative Health Impacts

Air pollution caused by the burning of fossil fuels was responsible for 8.7 million deaths globally in 2018. That's one in five people who died that year.

The Global Carbon Project estimates fossil fuel emissions will increase in 2021, reaching up to 36.4 gigatons of CO_2 ($GtCO_2$).

25 percent (roughly 9.1 $GtCO_2$) of that pollution is projected to come from the production of electricity and heat. The burning of coal, natural gas, and oil for electricity and heat is the largest single source of global greenhouse gas emissions.

There are documented links between pervasive air pollution from burning fossil fuels and cases of heart disease, respiratory ailments, and even the loss of eyesight.

A 2018 study showed that approximately 20 percent of all deaths worldwide were the result of illness from the pollution caused by the burning of fossil fuels.

⊕ **605**

GLOBAL DEATHS ATTRIBUTABLE TO AIR POLLUTION

Chronic obstructive pulmonary disease (COPD)	40%
Lower-respiratory infections (such as pneumonia)	30%
Stroke	26%
Ischemic heart disease	20%
Diabetes	20%
Neonatal deaths	20%
Lung cancer	19%

About 90 percent of the world's population breathes dangerously high levels of air pollution.

Human Health Impact: An Overview

In 2021 the World Health Organization called climate change "the single biggest health threat facing humanity." Today many people experience ill health as a result of factors like air pollution, extreme weather, food insecurity, disease, and stressors on mental health. Each year these environmental issues cause an estimated 12.6 million deaths. This means that nearly one in every four deaths globally is caused by environmental factors.

Rising levels of carbon dioxide in the atmosphere contribute to air pollution, temperature increases, rising sea level, and a growing number of extreme weather events like floods, wildfires, heatwaves, and droughts—all of which are direct causes of injuries and death. They might also increase the risk of mental health conditions like depression and anxiety.

Ecosystem change is also leading to decreased agricultural productivity, putting many people at risk of malnutrition. Directly and indirectly, climate change worsens a multitude of human health problems, including malaria and asthma.

⊕ **062**

DIFFICULT TO BREATHE

By 2050, outdoor air pollution will become the top cause of environmentally-related deaths worldwide.

Heat and Health

Over the last decade, extreme temperatures have killed more than 166,000 people worldwide. Rising ambient temperatures affect all populations, but some are more vulnerable, including older adults, infants and children, pregnant women, outdoor and manual laborers, athletes, and the poor.

Globally, 125 million people have been exposed to heatwaves during the last 15 years. Extended periods of high day and night temperatures increase stress on the body. Heat gain can result from a combination of external heat from the environment and internal heat caused by bodily functions. This rapid temperature rise affects the body's ability to control temperature and can create many health challenges.

Hospitalizations and deaths from extreme heat can occur the same day or several days after exposure. Temperature extremes can worsen chronic ailments, including cardiovascular, respiratory, and kidney diseases, as well as diabetes-related conditions.

⊕ **063**

Health impacts of exposure to extreme heat

Indirect Impacts

Accidents

Drowning
Work-related accidents
Injuries and poisonings

Transmission of

Food and waterborne diseases

Marine algal blooms

Disruption of Infrastructure

Power
Water
Transport
Productivity

Health services

More ambulance call-outs and slower response times

Increase number of hospital admission

Impact on storage of medicines

Direct Impacts

Heat Illness

Dehydration
Heat cramps
Heat stroke

Hospitalization

Stroke
Respiratory disease
Diabetes mellitus
Renal disease
Mental health conditions

Accelerated Death

Respiratory and cardio-vascular disease

Other chronic diseases (mental health conditions, renal disease)

Long-Term Wildfire Impacts: Effects of Smoke

Around the world, people have grown used to seeing pictures of skies turned orange by wildfires. But what is only now becoming clear is that the effects of wildfire smoke exposure can be long-lasting—particularly for children (and even those who were in their mother's womb when the smoke was present).

Wildfire smoke includes:

· carbon dioxide
· carbon monoxide
· nitrogen oxides
· volatile organic compounds (such as formaldehyde and benzene)

Of all the compounds in wildfire smoke, the most dangerous to human health are tiny bits of particulate matter known as PM2.5.

Exposure to these particles can create immediate effects like red itchy eyes, a sore throat, and wheezing. They can do longer-term damage too by being inhaled deep into the lungs.

Thirty of these particles end to end are as wide as one human hair. As a result, they are difficult for bodies to filter out.

> ## *I am glad I will not be young in a future without wilderness.*
>
> — Aldo Leopold

Of all the compounds in wildfire smoke, the most dangerous to human health are tiny particles known as PM2.5.

Research has found that exposure to PM2.5 may compromise lung development in children and aggravate or cause chronic lung diseases like asthma. Animal studies seem to indicate that microscopic particles may even be able to cross from the bloodstream into the tissues of the brain. Several studies have shown adverse effects on children's birth weight—and even their adult health status—from their mother's exposure to smoke while pregnant.

⊕ **085**

NEW OLD PATHOGENS

Scientists suspect that some of the abrupt thawing of permafrost may be reviving pathogens deadly to muskoxen, caribou, and nesting birds. Massive die-offs of muskoxen in Canada, as well as reindeer in Siberia, appear to be related to once-dormant pathogens that are being revived and released by rising temperatures.

Food and Waterborne Diarrheal Disease

More than 2,000 children die each day from diarrhea globally, accounting for more than 800,000 child deaths each year. Diarrhea is the second leading cause of child mortality in the world, and the leading cause of malnutrition for children under five years old.

Diarrheal disease:
- is most deadly for the elderly and children under five
- kills directly through dehydration
- kills indirectly by causing malnutrition, reducing the body's ability to ward off and recover from other diseases

Diarrheal disease and climate change

Most diarrheal disease is contracted through contaminated food and water sources. Research shows climate change is increasing its prevalence. It does this through three factors primarily:

1. Higher temperatures

Food spoils quicker in higher temperatures. Pathogens that cause diarrheal disease replicate faster and live longer. An increase of just 1°C/1.8°F in temperature results in a 3.8 percent increase in childhood clinic visits for diarrhea.

2. Increased rain events

Heavy rains wash bacterial and viral contamination from areas with poor sanitation into water supplies. This increases exposure to diarrheal disease. For example, moderate to strong El Niño events increased diarrhea in children by four percent.

3. Increased drought

When droughts occur, people are forced to use less healthy sources of water. This increases the likelihood of drinking, washing, watering, or preparing food in contaminated water. In one study conducted in Peru, the dry season led to a 1.4 percent increase in diarrhea.

🌐 589

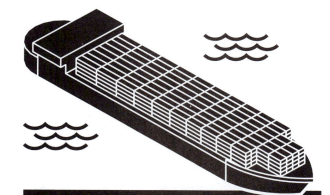

WATERBORNE EMISSIONS

In 2015, research estimated that emissions generated by the world's 90,000 cargo ships were responsible for 60,000 deaths from pollution.

Just one of the world's largest container ships is capable of emitting as much pollution as 50 million cars.

The world's 15 largest ships pump out enough emissions to equal the pollution caused by all of the 760 million cars on the planet.

The nitrogen oxide and sulfur oxide released are known causes of cancer and asthma.

The low-grade fuel used in shipping contains 2000 times as much sulfur as diesel fuel for trucks and cars.

DIARRHEA AND CHILD MORTALITY

Diarrhea is the second leading cause of child mortality in the world and the leading cause of malnutrition for children under five.

The Impact of Global Tourism

Access to lower-cost car and air travel has had a dramatic impact on the number of tourists worldwide. In addition to travel, tourism uses energy for lodging, service, and shopping.

Tourism's carbon footprint accounts for 8 percent of man-made global carbon emissions. Almost half of those emissions are created by transportation. For domestic tourism, residents within their own countries travel primarily by car, followed by air flight. International travel is mainly conducted by air, both within the same region and across regions.

There were 25 million tourist arrivals by air in 1950. Sixty-eight years later, in 2018, the number of tourists traveling by air was 1.4 billion. That is a 56-fold increase.

In 2030, it's projected that greenhouse gases will increase by 25 percent, with two gigatons of transportation-related CO_2 attributable to tourism.

🌐 **072**

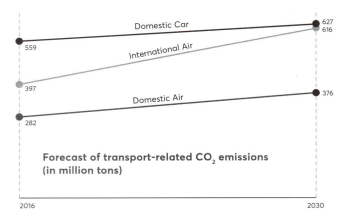

Forecast of transport-related CO_2 emissions (in million tons)

Domestic Car — 559 ... 627
International Air — 616
397
Domestic Air — 282 ... 376

2016   2030

The Costs of Sea Level Rise

Sea-level rise (SLR) is likely to have numerous negative economic impacts, primarily on coastal cities around the world. 44 percent of the world's population live within 150 kilometers of the coast.

44 percent of the world's population live within 150 kilometers of the coast.

The economic impacts of SLR include:
- Flooding damage that could cost $14 trillion worldwide annually by 2100.
- Expected direct annual losses due to coastal floods (without adaptation) that could amount to 0.3-9.3 percent of global GDP by 2100.
- Ports, power plants, transmission lines, oil refineries, sewage treatment facilities, telecommunications cables, and highways have all been built close to coastlines. The cost to build vital infrastructure to defend against SLR over the next 20 years will be greater than $400 billion in the US alone.

🌐 **603**

Carbon emissions from global tourism

49% TRANSPORT
12% GOODS
10% F&B
29% OTHERS

Impact of Carbon Exports & Imports

importers		
−352	USA	
−180	Japan	
−158	UK	
−121	Italy	
−117	France	
−106	Germany	
−84	Switzerland	
−82	Belgium	
−73	Hong Kong	
−66	Singapore	
−51	South Korea	
−51	United Arab Emirates	
−48	Mexico	
−44	Brazil	
−33	Turkey	
−33	Spain	
−29	Vietnam	
−29	Sweden	
−26	Canada	
−26	Philippines	
−26	Austria	
−22	Indonesia	
Qatar	22	
Ukraine	26	
Iran	48	
South Africa	139	
India	195	
Russia	257	
China	1,369	

exporters

Picking up a Costa Rican banana at the supermarket and then calling home on a smartphone made in China is possible because of international trade.

When goods and services move between the country of production and the country of consumption, carbon emissions move with them.

Approximately 25 percent of all CO_2 emissions from human activity 'flow' from one country to another via imports and exports. As to be expected, these flows are not evenly distributed across products or countries.

Commodities such as steel, cement, and chemicals account for roughly half of all cross-border carbon flows. The remaining half is contained in semi-finished/finished products like cars, clothing, industrial machinery and equipment. Bananas (it turns out) don't make up a significant amount of world-wide carbon flows.

In 2014, the US imported 352 Mt of CO_2 and China exported 1,369 Mt CO_2.

The wealthier countries, because of their cross-border carbon imports, are pushing the dirty job of combustion to other countries. Tracking this flow will make it easier for governments to properly account for and limit carbon emissions.

⊕ **578**

Importers and exporters of CO_2
(in megatons)

Impact of Commercial Space Travel

While the carbon impact of commercial space travel is still relatively low, its growth potential is large.

In 2020, there were 114 launches related to commercial space travel. This number is estimated to increase to 360 per year by 2030 and to 1,000 per year in the future. Virgin Galactic's CEO hopes to launch at least 400 flights from each spaceport annually.

Space travel has a large carbon footprint due to the burning of fossil fuels. Rocket engines produce soot, which consists of tiny carbon particles that are so light they can stay in the upper reaches of the atmosphere for years. These pollute the stratosphere (from 10 km above Earth's surface and upwards) and the mesosphere (50 km upwards). Soot absorbs ultraviolet light, which means that it could heat the stratosphere.

Along with greenhouse gases, high-grade kerosene-fueled rockets emit chlorine and aluminum oxide that destroy the ozone layer. VSS Unity, part of the Virgin Galactic fleet, uses a hybrid fuel made of a solid carbon-based fuel, hydroxyl-terminated polybutadiene (HTPB), and nitrous oxide. The SpaceX Falcon series of reusable rockets will launch the Crew Dragon into orbit using liquid kerosene and liquid oxygen. It would take an average gas-powered car more than two hundred years to emit as much carbon dioxide as SpaceX's Falcon Heavy does in just a few minutes.

⊕ 082

CO_2 emissions for each of the passengers on a commercial space flight would be between 50 and 100 times more than the 1-3 tons of emissions that are generated per passenger on a long-haul airplane flight.

Eco-Anxiety

Climate change is having an impact on mental health. While eco-anxiety is not yet a diagnosable condition, it is pervasive. Recent surveys indicate close to 70 percent of all adults and 85 percent of all children experience some form of eco-anxiety.

Sarah Niblock of the UK Council for Psychotherapy shared that "eco-anxiety is a perfectly normal and healthy reaction" to the climate change threat. In short, eco-anxiety is an emotional signal to pay attention and to act. It is a stress response and thus a survival mechanism designed to prime individuals to respond.

What's the best response?
Guidance from the American Psychiatric Association (APA), Yale University's Climate Connection Initiative, and psychologists worldwide can be summarized into an action plan to address eco-anxiety.

These steps can create a greater sense of individual security and agency in the face of eco-anxiety and can help build the resiliency needed for the challenges ahead:

1. Acknowledge that it's okay to feel this way
Experiencing eco-anxiety is perfectly normal. It is something that others are feeling too. Practice self-compassion and be kind and understanding toward yourself by acknowledging this fact.

2. Break the "spiral of silence" by talking
While the majority of the population is experiencing eco-anxiety, upwards of 64 percent of adults in the US never or only rarely discuss climate change. There is a problematic "spiral of silence" (as researchers call it) related to climate change and eco-anxiety. By openly discussing eco-anxiety, the feelings associated with it can be normalized and better managed.

Since conversations about climate change and feelings can be difficult, consider an "active listening" strategy. Questions such as "What have you noticed on the news about climate change that really struck you?" or even, "What are you feeling? What's on your mind?" are helpful conversation starters to pair with active listening.

3. Practice self-care
Psychologists recommend three strategies to combat the negative effects of stress in healthy ways:
· improved sleep
· moving more
· eating healthy
Sleep is a very good place to start as lack of it may have the biggest impact on the ability to manage daytime stress. More than a third of American adults are not getting enough sleep on a regular basis, so it is likely that those with eco-anxiety are not getting enough sleep. Exercise (even a regular brisk walk) can improve sleep and directly combat stress. Indeed, moderate physical activity has been shown to cut stress levels in half. Finally, a varied and nutritious diet can provide more physical energy to deal with eco-anxiety related challenges.

4. Strengthen your network of support
Success in discussing feelings—and successfully practicing self-care—is in part predicated on establishing solid relationships. Whether at the immediate in-person and local level or online and virtual, strive to build strong social networks with family, friends, neighbors, and others that are built on trust and regular engagement.

5. Take action to positively address the climate change challenge

Finally, know that taking preemptive action provides mental health benefits. Besides reading (and sharing) this *Almanac*, there are many other actions you can take. Your actions can serve to limit climate change impacts globally over the long-term as well as to prepare for it locally in the short term. When doing so, pace yourself. Engage at an effort level that feels comfortable and that you believe you can sustain over time.

Seek out professional help when you need it

If you or a person you care about is finding that eco-anxiety is significantly interfering with everyday life activities, the ability to work, or the ability to be safe, please seek professional help. If the concern is immediate and significant dial 1-800-273-8255 for the US National Suicide Prevention Lifeline or text TALK to 741741 for free, anonymous 24/7 crisis support in the US from the Crisis Text Line. If outside the US, find other international suicide helplines at Befrienders Worldwide.

🌐 **252**

JOINING HANDS

Since 1973, Antaryami Sahoo, a teacher in rural Odisha, India, has planted 10,000 trees in public spaces. He has led his students in planting an additional 20,000 trees. In his words, "Joining hands can achieve wonders."

We need to change our way of thinking and seeing things. We need to realize that the Earth is not just our environment. The Earth is not something outside of us.

Breathing with mindfulness and contemplating your body, you realize that you are the Earth. You realize that your consciousness is also the consciousness of the Earth. Look around you—what you see is not your environment, it is you.

— Thich Naht Hanh

Solutions

Creating the world we want

The Drawdown Rankings

A worldwide team of experts led by Paul Hawken ranked hundreds of possible solutions to the challenge of climate change. Forty-nine of the most impactful global solutions are listed here. For more information, visit drawdown.org. Scenario numbers represent cumulative gigatons of CO_2 reductions between now and 2100.

Scenario 1 is based on a 2°C temperature rise by 2100. Scenario 2 is based on a 1.5°C temperature rise by 2100. Solutions may differ with context and economic, ecological, social, and political conditions of a given country.

🌐 245

SOLUTION	SECTOR(s)	SCENARIO 1	SCENARIO 2
Reduced Food Waste	Food, Agriculture, and Land Use	90.70	101.71
Health and Education	Health and Education	85.42	85.42
Plant-Rich Diets	Food, Agriculture, and Land Use	65.01	91.72
Refrigerant Management	Industry / Buildings	57.75	57.75
Tropical Forest Restoration	Land Sinks	54.45	85.14
Onshore Wind Turbines	Electricity	47.21	147.72
Alternative Refrigerants	Industry / Buildings	43.53	50.53
Utility-Scale Solar Photovoltaics	Electricity	42.32	119.13
Improved Clean Cookstoves	Buildings	31.34	72.65
Distributed Solar Photovoltaics	Electricity	27.98	68.64
Silvopasture	Land Sinks	26.58	42.31
Peatland Protection and Rewetting	Food, Agriculture, and Land Use	26.03	41.93
Tree Plantations (on degraded land)	Land Sinks	22.24	35.94
Temperate Forest Restoration	Land Sinks	19.42	27.85
Concentrated Solar Power	Electricity	18.60	23.96
Insulation	Electricity / Buildings	16.97	19.01
Managed Grazing	Land Sinks	16.42	26.01
LED Lighting	Electricity	16.07	17.53
Perennial Staple Crops	Land Sinks	15.45	31.26
Tree Intercropping	Land Sinks	15.03	24.40
Regenerative Annual Cropping	Food, Agriculture, and Land Use	14.52	22.27
Conservation Agriculture	Food, Agriculture, and Land Use	13.40	9.43
Abandoned Farmland Restoration	Land Sinks	12.48	20.32
Electric Cars	Transportation	11.87	15.68

Multistrata Agroforestry	Land Sinks	11.30	20.40
Offshore Wind Turbines	Electricity	10.44	11.42
High-Performance Glass	Electricity / Buildings	10.04	12.63
Methane Digesters	Electricity / Industry	9.83	6.18
Improved Rice Production	Food, Agriculture, and Land Use	9.44	13.82
Indigenous Peoples' Forest Tenure	Food, Agriculture, and Land Use	8.69	12.93
Bamboo Production	Land Sinks	8.27	21.31
Alternative Cement	Industry	7.98	16.1
Hybrid Cars	Transportation	7.89	4.63
Carpooling	Transportation	7.70	4.17
Public Transit	Transportation	7.53	23.40
Smart Thermostats	Electricity / Buildings	6.99	7.40
Building Automation Systems	Electricity / Buildings	6.47	10.48
District Heating	Electricity / Buildings	6.28	9.85
Efficient Aviation	Transportation	6.27	9.18
Geothermal Power	Electricity	6.19	9.85
Forest Protection	Food, Agriculture, and Land Use	5.52	8.75
Recycling	Industry	5.50	6.02
Biogas for Cooking	Buildings	4.65	9.70
Efficient Trucks	Transportation	4.61	9.71
Efficient Ocean Shipping	Transportation	4.40	6.30
High-Efficiency Heat Pumps	Electricity / Buildings	4.16	9.29
Perennial Biomass Production	Land Sinks	4.00	7.04
Solar Hot Water	Electricity / Buildings	3.59	14.29
Grassland Protection	Food, Agriculture, and Land Use	3.35	4.25

SHEEP'S WOOL AS PACKAGING

Over 200,000 tons of sheep's wool is discarded annually in Europe. An unused resource for sustainable and protective packaging, this same wool could meet 120 percent of the global demand for bubble wrap.

Greenwashing & Recycling Theatre

Many towns and organizations have dramatically increased the effectiveness of their recycling programs. But as marketers become more aware of consumers' environmental concerns, some corporations can misrepresent the impact of their sustainability efforts. Not all green practices and recycling claims offer an honest picture.

Industries that benefit from producing difficult-to-recycle items like plastic often mislead the public about the effectiveness of recycling. This is amplified by the recycling industry itself, a large and growing sector of the economy.

Some organizations have had great success in decreasing the impact of the goods and services they provide. However, corporations that hide harmful practices behind the appearance of environmental stewardship are practicing "greenwashing."

Recycling

In 2018, the US EPA reports that the total Municipal Solid Waste (MSW) recycled was more than 63 million metric tons. Percentages recycled have increased since 1960, with 68 percent of paper and paperboard recycled in 2018 compared to 28 percent in 1990 and 25 percent of glass in 2018 compared to 20 percent in 1990.

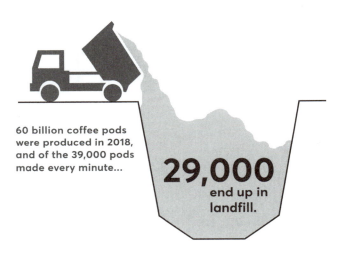

60 billion coffee pods were produced in 2018, and of the 39,000 pods made every minute...

29,000 end up in landfill.

Percentage of waste recycled by material type

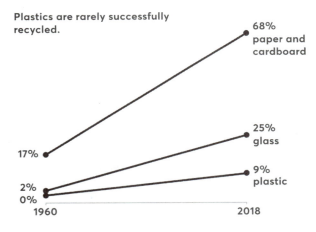

Plastics are rarely successfully recycled.

- 68% paper and cardboard
- 25% glass
- 9% plastic
- 17%
- 2%
- 0%
- 1960
- 2018

The recycling and reuse industry provides more than 681,000 jobs and generates more than $37.8 billion in wages. The largest US company, Waste Management, grosses more than $15 billion in annual revenue. The major players in the recycling industry are public companies.

Greenwashing

Here's a specific example of how an activity can appear to be sustainable while its net impact remains negative. Consider the trend toward single-use coffee and tea pod systems in homes and offices.

The transition from sharing office coffee pots, with their biodegradable filters, to using tiny disposable plastic pods has generated a large-scale impact. Halo, a sustainable coffee company, estimates that 60 billion coffee pods were produced in 2018, and of the 39,000 pods made every minute, 29,000 end up in landfills.

Some of the pod companies have launched voluntary recycling programs, but there's no evidence that they have had much effect, and the programs ignore the broader question of whether such products should exist at all.

> **While plastic recycling has been promoted as an environmentally friendly activity, it's not. Less than 10 percent of plastic is recycled, regardless of how much is put in the recycling bin.**

What happens to plastic when it's put in for recycling at designated collection points? An estimated 31 percent of waste ends up in landfills and the majority is incinerated.

There are many kinds of plastic, and they're cumbersome or impossible to sort. Even when perfectly sorted, most plastic can't be recycled. The types that can eventually degrade after a few generations and then must be discarded.

When mixed plastics are deposited in a recycling bin it causes 'contamination' and the entire bin is often incinerated. These incinerators are often called "waste-to-energy" plants and promoted as a sustainable trash disposal alternative. In fact, they're simply burning fossil fuels that had a short life as a plastic bag or bottle first.

Municipal incinerators are eager to take in plastic as waste, also called "feedstock." It's a cheap way to produce heat. Electricity produced by these plants is often incorrectly labeled as sustainable or "green" energy. In fact, municipal incinerators produce large amounts of greenhouse gases—more than coal-fired powered stations.

In one US EPA study, electricity generated by incineration produced 1.36 metric tons of CO_2 per kWh, while coal-fired generation produced 1.02 metric tons per kWh.

> **In 2019, the production and incineration of plastic produced the same amount of greenhouse gases as 189 500-megawatt coal power plants.**

⊕ **089**

Workforce size across various industry (in millions)

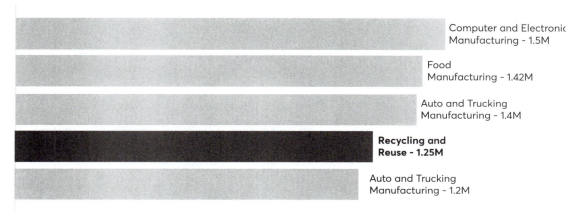

Computer and Electronic Manufacturing - 1.5M

Food Manufacturing - 1.42M

Auto and Trucking Manufacturing - 1.4M

Recycling and Reuse - 1.25M

Auto and Trucking Manufacturing - 1.2M

The recycling industry employs more than one million people in the US.

Bioplastics

Plastics have impacts in their creation (carbon is emitted) and disposal (landfills or incineration).

Plastics are traditionally made from long polymer chain molecules derived from fossil fuels. The polymer chain doesn't break down easily, which means plastics are non-biodegradable.

Long polymer chain molecules exist elsewhere in nature, however. These include polysaccharides (such as starch and cellulose), proteins (gluten, gelatin), and lipids (fats, oils). These polymers can be made into plastics.

Bioplastics are also biodegradable. The timing and method of degradability varies. Some require an enzyme to break down; others require high temperatures. Some bioplastics degrade in water.

How are bioplastics more sustainable than traditional plastics?
- The use renewable raw materials.
- They're often biodegradable.
- They're less toxic.

Increased production of bioplastics comes with several drawbacks. Producing biomass on a large scale competes with food production for land and water resources and can require fossil fuel-based fertilizers.

🌐 **256**

Fast Fashion and Carbon

Around 1890, Levi Strauss began selling the 501 line of jeans. They still do.

Most fashion doesn't work this way. Styles change quickly, and producers race to bring new ideas to market to keep up with demand. Between 2000 and 2014, worldwide consumers bought 60 percent more clothing per capita, and they're keeping clothing for half as long. Some items are worn only seven or eight times.

The fashion industry produces more items, and the cycle of buying and discarding continues.

Garments are difficult to recycle. According to McKinsey & Co., "For every five garments produced, the equivalent of three end up in a landfill or incinerated each year."

In addition to exporting millions of pounds of fabric and clothing waste to countries like Chile, the fashion industry is increasingly relying on polyester, which is created from fossil fuels.

Global garment production has more than doubled since 2000. Some fast fashion brands release collections *weekly*, replacing the two traditional fashion seasons of Fall/Winter and Spring/Summer.

This cycle of production and waste leads to the release of 4 percent of all global greenhouse gas emissions, matching the entire economies of France, Germany, and the United Kingdom combined. Emissions include manufacturing, shipping, and incinerating unwanted clothing.

🌐 **101**

A Carbon Dividend and Fee

Market-driven economies respond to market forces. And the two most direct forces are taxes and payments. People buy what's on sale and avoid what's expensive.

At the same time, coordinating policy-making between countries is challenging, especially when it comes to climate.

If a country implements a carbon tax, it places a financial burden on its citizens and reduces the competitiveness of domestic industries compared to those of other, non-taxing countries. If a home industry bears the added costs of a carbon tax, it likely will lose business to foreign firms that don't face such a tax.

Faced with a carbon tax, it's expected that companies will either innovate or emigrate. If innovation is seen as too costly, companies might move production abroad, which has occurred in some industries when labor protections were put into place.

This possibility makes countries hesitant to impose carbon taxes; they want to avoid the loss of industry to non-taxing nations, a phenomenon known as "carbon leakage."

Border carbon adjustment

An alternative being considered is a border carbon adjustment. Instead of requiring multi-lateral agreements, this approach seeks to lower carbon use and also solve carbon leakage. Under this plan, countries with carbon taxes would impose a tariff on foreign products that were not taxed on carbon where they were produced. As a result, domestic and foreign goods would bear the same carbon fee, incentivizing industries across-the-board to reduce carbon use. The message of such a fee is, "If you want to sell in this country, you need to lower your carbon footprint, no matter where you produce."

Carbon pricing

Historically, coal and fossil fuels were used because they were relatively inexpensive to produce, regardless of the cost to human health and the environment.

Carbon pricing seeks to build in health and environmental costs from the start, so that market forces will reward wise choices about what to consume and combust.

Devices such as carbon fees and taxes, "cap-and-trade," and carbon offsets help create economic incentives that spur a faster and more innovative transition away from carbon. They impose an additional financial cost onto emissions while allowing industries to continue functioning as they shift to less carbon-intensive operations.

Carbon dividend

But what happens to the money collected through a fee or a tax? One plan is to pay a dividend, a check written directly to every household. The Climate Leadership Council Carbon Dividend Plan proposes a tax rate in 2023 of $44/per ton of CO_2 emitted through combustion, with annual payments made to every family (estimated to be $2,000). By 2035 the tax would rise to $79 a ton and the payments to families increase as well.

All net proceeds from the carbon fee would be returned to individuals on an equal and quarterly basis as a "carbon dividend." The average person will get back about as much as they pay in increased energy costs. Profligate carbon users will pay more, and consumers who spend carefully will benefit. An analysis by Resources for the Future finds that this simple approach would reduce emissions levels by 27 percent in the US over 12 years.

Cap and trade

A "cap-and-trade" approach offers an alternative to a carbon tax. A jurisdiction sets a maximum level of emissions and issues permits to companies. Companies that exceed their permitted amount pay penalties. Companies can also trade unused permitted amounts to other companies. The theory is that the permits will reflect the true value of the carbon, and as soon as a cheaper way to produce the same goods is discovered, companies will switch over. The total amount of permits declines each year, creating steeper penalties for emissions over time. Notable adopters of this approach include the state of California and the European Union.

🌐 **239**

Mass Transit

Mass transit makes cities more energy efficient than suburbs or rural areas.

Coordinated effort can move more people with less energy. According to the United States Bureau of Labor Statistics, the average American spends almost 16 percent of their budget on transportation costs.

Luxembourg made all public transportation free as of March 2020, and some American cities are following suit, offering free mass transit to entice new riders.

🌐 **246**

🌐 **246**

ONE-QUARTER OF LOS ANGELES

Parking takes up 14 percent of the total area of Los Angeles. Another 10 percent is freeways and roads.

50 people walking

50 people riding bikes

50 people on a bus

50 people in 36 cars

```
G O G I V E R Y T I R U C E S N I Q D R A W D O W N Y V T N
X E S W I J J Y G R E N E Y C F R L N F G A F L O S O R E A
H U U E Q U I V A L E N T T C O U J F N N N L B I O M A S S
U Y V A G R E E N Y O I E C F R U O I U B M I B I O F U E L
O W H T N E D K O I S G I F V S R H U K T E F L F S J Q E F
S B Y H S Z E E T R D R S V T E S L W T E S E L D C Y X M E
N F D E Z Y L A E U T E A I S A O H E E M U N I I L C P E D
O F R R T C R V B C T X C T W C Y N A G I O O F O E N O R I
I Y O C I G I U E T U E A N O D E T N A S H I D R A E C T X
S G G H I D T L E H R T E T R C L I V R S N T N E N I A X O
O Z E M O N E E H W I E O O O A M G K O I E A A T F C B E I
R V N I E U H A A O R R E P I R E N N T O E I L S V I F A D
E Z B V D S T S N G P L O R A O U O V S N R D D A O F P N G
Z Z E V I E T D J T E R T W E Z I C R G H G A G P G F O D O
K N O W L E D G E C H S D N Y T E I X N A O R L G K E R F S
B P C F O X G T T U R G E I T A N E U T R A L L D H C B X
R O A L D E M R N D D I O S A C I R E M A S E G A K N I S T
X U P U E O I A N L N N N P S D E S E R T I F I C A T I O N
G S T O B C O I E E Y A N P O X T E Z I P H G C I T Y C H I
E A U R L C S I E Q R P G G W G C K C Z G F T R E R T O D R
O C R I A B Y R W T G A S E S G E S F F H B X I R A I E F P
T T E N N O I T U L O V E R S U O N E G I D N I W D R C O T
H I C A P N Y B W S S C Y C L E R D I P L A N D D I U O S O
E V L T G S Q D N B J D E R A R F N I C L O C K I N C S S O
R I I E R E P O R T I N G C C P I A C T I V I S T G E Y I F
M S M D G G T G X G J S E Q U E S T R A T I O N Z U S S L N
A M A X R A L L O D A Y X N O I T A T S E R O F E D Y T M D
L V T E G T S E R O F S H N O I T A Z I L I T U E N P E E O
E H E I L O S S C H A N G E C U E P O H G L O B A L A M C O
Y C G R E W O P P O T E N T I A L S N O I S S I M E M B I F
```

FIND THE WORD LIST AT 🌐 777

The Rise of E-Bikes

Electric bicycles first arrived in the 1890s(!), with the first modern pedal-assist e-bike marketed by Yamaha in 1993.

Modern e-bikes are a great way to lower the carbon intensity of transportation. Including manufacturing and shipping, their carbon impact is still far lower than any electric car or truck.

Beyond carbon advantages, bicycles make more efficient use of existing infrastructure:
· they are well-adapted to today's roads;
· require far less road space than cars;
· create no tailpipe or noise pollution;
  · can be purchased for a fraction of the price of a car (even for high-end e-bikes).

Electric bicycles significantly rose in popularity during 2020 and 2021 as the COVID-19 pandemic changed lifestyles around the world. Sales during this period grew by 240 percent.

With the same emissions, an e-bike can go 96 times farther than a car.

More than half of all trips taken by car in the United States are shorter than 16 km. A car that travels 64 km per day will emit 7000kg of CO_2, while an e-bike that travels the same distance will emit 300kg of CO_2—about 96 percent less.
🌐 **234**

Electric Vehicles

Electric vehicles (EVs) are battery-operated and run on an electric motor instead of a fossil fuel-fired internal combustion engine (ICE). EVs do not emit tailpipe emissions and are classified as zero-emission vehicles. Hybrid EVs, by contrast, have both an ICE and an electric motor.

EVs are charged by plugging large battery packs into a public charging station or a home garage wall outlet. They are also highly efficient. EVs convert 77 percent of electrical energy to power at the wheels, compared with ICE engines, which convert only 12 to 30 percent.

Globally, there are more than 12 million passenger EVs and one million commercial EVs (i.e., trucks) on the roads. Since 2015, the global share of new passenger EVs has increased roughly 50 percent per year. 3.1 million new passenger EVs were sold in 2020. That's up 67 percent from just the year before.

Projections for EV adoption are robust:

- by 2025, 15 million in annual sales of EVs globally
- by 2038, ICE car sales will peak as the world shifts to electric vehicles
- by 2040, 70 percent of new vehicles globally will be EVs

⊕ **100**

States with the highest rate of EV charging ports per 100,000 people

STATE	EV PORTS PER 100,000 PEOPLE
Vermont	125.8
Washington, D.C.	88.1
California	82.0
Hawaii	52.5
Colorado	52.2

States with the largest growth of EV charging ports per 100,000 people

STATE	2021 Q1 GROWTH OF EV PORTS
Oklahoma	52.3%
North Dakota	16.7%
Michigan	10.8%
Pennsylvania	10.5%
Massachusetts	9.7%

Electric cars

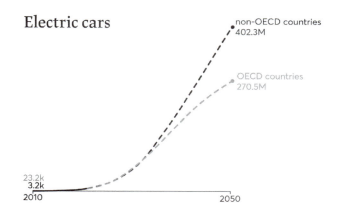

non-OECD countries 402.3M

OECD countries 270.5M

23.2k
3.2k
2010   2050

Transportation (including commuting) is the second largest source of greenhouse gas emissions in the United States.

How Roundabouts Help Lower Emissions

A roundabout (sometimes known as a traffic circle or rotary) is a circular intersection where traffic flows in one direction around an island. There are no stop lights involved.

> **Roundabouts are designed to reduce collisions and traffic jams. They also help to lower carbon emissions by reducing fuel consumption.**

When roundabouts are installed instead of traffic signals or stop signs, carbon monoxide emissions are reduced by 15-45 percent, nitrous oxide emissions by 21-44 percent, carbon dioxide emissions by 23-34 percent, and hydrocarbon emissions up to 40 percent. Overall, fuel consumption is reduced by an estimated 23-34 percent.

Based on a 2005 study, the Insurance Institute for Highway Safety estimates that if 10 percent of the intersections in the United States were converted from traffic lights to roundabouts, vehicle delays in 2018 would have been reduced by more than 981 million hours and fuel consumption by over 2,476 million liters.

Based on a study done in Virginia, Mike McBride, the former city engineer of Carmel, Indiana, estimates that one roundabout in his city of about 100,000 people saves 75,708 liters of gas annually.

The magic roundabout

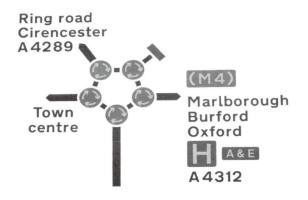

In addition to the automobile-related benefits of roundabouts, these intersections require no electricity, making them more sustainable than intersections regulated by traffic lights.

The two biggest disadvantages are that they require more land and that drivers are initially opposed to change.

However, as drivers become more familiar with these types of intersections, they become more supportive. In one study, public support for two roundabouts in Washington State went from 34 percent prior to construction to 70 percent once people had a chance to get used to the new road feature.

⊕ **230**

Energy-Efficient Cars

Transportation is responsible for 24 percent of CO_2 emissions from energy worldwide. About 45 percent comes from passenger vehicles like cars, motorcycles, and buses. There are thousands of varieties of cars available, so in 2021 the American Council for an Energy-Efficient Economy (ACEEE) ranked the twelve models that have the least impact on the environment over their lifecycle.

The list considers:
- raw materials used to construct the vehicle
- emissions associated with the manufacturing and use of the automobile
- effects of recycling or disposing of the vehicle once it is no longer usable

The ACEEE focused particularly on tailpipe emissions, fuel economy, vehicle mass, and battery mass and composition. Pollutants are assigned a cost. When all this information is plugged into the formula ACEEE devised, each vehicle earns its *environmental damage index* or EDX.

No car was awarded a 100 (or a zero EDX) for its "Green Score" on the list below. All cars included on the list are electric vehicles, of which there are three types: EVs, HEVs, and PHEVs.

- EVs are electric vehicles that run on battery power alone. Since they lack an internal combustion engine, they can go further distances on a single charge than HEVs.
- HEVs are hybrid electric vehicles that run on an internal combustion engine and a battery that stores energy in an electric motor. These batteries charge through "regenerative braking," where the act of braking stores kinetic energy and decreases the amount of fuel the internal combustion engine burns.
- PHEVs are plug-in hybrids with an internal combustion engine and a battery-powered electric motor. These vehicles can store enough battery energy to travel up to 40 miles at a time and can decrease fuel consumption by up to 60 percent.

🌐 **226**

The twelve most efficient cars for 2021

According to the American Council for an Energy-Efficient Economy

1. Hyundai Ioniq Electric EV; Green Score: 70
2. MINI Cooper SE Hardtop EV; Green Score: 70
3. Toyota Prius Prime PHEV; Green Score: 68
4. BMW i3s EV; Green Score: 68
5. Nissan Leaf EV; Green Score: 68
6. Honda Clarity PHEV: Green Score: 66
7. Hyundai Kona Electric EV: Green Score: 66
8. Kia Soul Electric EV; Green Score: 65
9. Tesla Model 3 Standard Range Plus EV; Green Score: 64
10. Toyota RAV4 Prime PHEV; Green Score: 64
11. Toyota Corolla Hybrid HEV; Green Score: 64
12. Honda Insight HEV; Green Score: 63

HYUNDAI IONIQ ELECTRIC EV

MINI COOPER SE HARDTOP EV

TOYOTA PRIUS PRIME PHEV

BMW I3S EV

NISSAN LEAF EV

HONDA CLARITY PHEV

HYUNDAI KONA ELECTRIC EV

KIA SOUL ELECTRIC EV

TESLA MODEL 3 STANDARD RANGE PLUS EV

TOYOTA RAV4 PRIME PHEV

TOYOTA COROLLA HYBRID HEV

HONDA INSIGHT HEV

The Changing Cost of Power

For years, the cost of installing renewable power was subsidized by governments. As technology and engineering have improved, the costs of these new approaches have dropped dramatically.

Communities building new power sources are increasingly choosing to build wind and solar operations. In many situations, wind and solar can now produce cheaper electricity than coal or natural gas. Hydroelectric remains a safe and cheap alternative, but there are few suitable sites available.

The costs of solar and wind energy have decreased over the past decade due to advancements in technology and economies of scale. The current starting cost (in 2021) of unsubsidized power operations are as follows:

- **wind:** $26/megawatt-hour
- **wolar:** $28/megawatt-hour
- **high-efficiency natural gas** (aka natural gas combined cycle): $45/megawatt-hour
- **coal:** $65/megawatt-hour

These megawatt-hour dollar amounts are calculated using the levelized cost of energy (LCOE) for each type of electricity production. LCOE divides the lifetime cost of a source of power by the energy it produces.

These projects operate over several decades. Wind and solar operations require occasional maintenance, while gas and coal operation costs fluctuate due to the cost of their daily fuel inputs changing over time.

⊕ 237

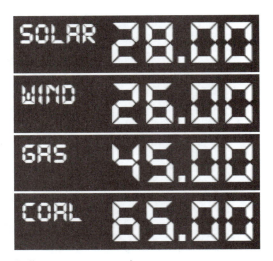

Dollars per megawatt-hour

HUGE GLACIER MELTING QUICKLY

In Antarctica, the Thwaites Glacier is the size of Great Britain. In the 1990s, 10 billion tons of ice melted from its edges each year. In 2020, that annual loss was eight times faster, as warmer ocean water moves under its eastern shelf. The entire glacier is now in danger of breaking apart, which could lead to its complete collapse. If it melts, this single glacier could raise the level of all the oceans in the world as much as 65 cm (more than two feet).

Energy Payback for Renewables

Carbon-based fuels are cheap, because they are under-priced, failing to include the environmental costs borne by everyone. But the low price of fossil fuel means that building a coal or gas-fired powerplant has a short payback time. The cost of building and maintaining the plant is quickly recouped by selling what it produces.

To build renewable facilities, investors want to understand how long it will take to earn back their investment. But in the case of renewables, in addition to evaluating the financial payback, it's worth considering the energy payback. A plant built on fossil fuels never achieves energy payback—it's carbon negative from the first day and gets worse over time.

Renewable energies aren't free. Solar panels must be built or turbines must be installed. Energy (and often non-renewable energy) is required to build renewable energy infrastructure like hydropower dams. In addition, energy is needed to decommission renewable energy infrastructure when it's too old to still be in use.

The table below outlines how long it takes for the energy used to create a plant to go from negative (carbon emitted) to positive (carbon saved compared to the alternatives) based on a 25-year lifetime of use.

⊕ **232**

Energy payback time over 25 years (300 months)

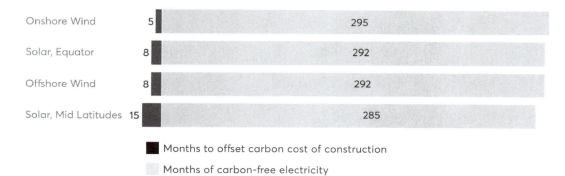

	Months to offset carbon cost of construction	Months of carbon-free electricity
Onshore Wind	5	295
Solar, Equator	8	292
Offshore Wind	8	292
Solar, Mid Latitudes	15	285

■ Months to offset carbon cost of construction

▪ Months of carbon-free electricity

Author James D. Newton described a conversation between inventor and entrepreneur Thomas Edison, automobile manufacturer Henry Ford, and tire manufacturer Harvey Firestone...

"We are like tenant farmers, chopping down the fence around our house for fuel, when we should be using nature's inexhaustible sources of energy—sun, wind, and tide," Edison said.

Firestone pointed out that oil and coal and wood couldn't last forever.

Edison responded, "I'd put my money on the sun and solar energy. What a source of power! I hope we don't have to wait till oil and coal run out before we tackle that. I wish I had more years left!"

Wind Energy

Wind as a source of energy has been around for thousands of years.

5,000 BC: Wind power propels boats along the Nile River.

200 BC: Wind-powered water pumps appear in China. Persia and the Middle East harness wind to grind grain.

1000 AD: Humans in the Middle East use wind pumps and windmills for food production.

1200 AD: The Dutch develop large windmills to drain lakes and marshes.

1700 AD: American colonists use windmills to grind grain, pump water, and cut wood at sawmills.

1800 AD: Homesteaders and ranchers install thousands of wind pumps in the settling of the western United States.

Late 1800s and early 1900s: Small wind-electric generators (wind turbines) are widely adopted.

A new age for energy production

Today, large wind turbines—offshore and on land—are used to generate electricity. A single, modern-day offshore wind turbine generates more than 6 megawatts (MW) of energy, enough to power thousands of homes.

On land-based wind farms, wind turbines generate on average one to five MW each.

Progression of wind turbine sizes and their energy output

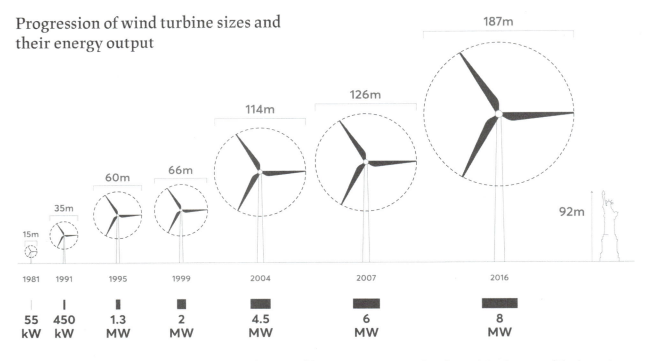

1981	1991	1995	1999	2004	2007	2016
55 kW	450 kW	1.3 MW	2 MW	4.5 MW	6 MW	8 MW

Wind power is one of the lowest-cost large-scale renewable energy sources on the planet. It is also one of the largest sources of renewable energy in the United States, surpassing that of hydroelectric in 2021.

At first, I thought I was fighting to save rubber trees.
Then I thought I was fighting to save the Amazon rainforest.
Now I realize I am fighting for humanity.

— Chico Mendes

How wind turbines work

A wind turbine is an electric motor in reverse. Instead of applying electrical energy to spin the motor, the force of the wind turns the motor's shaft and generates electrical energy. As the wind blows, the massive propeller-like blades of the turbine spin at between 13 and 20 revolutions per minute.

Wind speeds of approximately 15 km per hour or more are needed to generate electricity at most utility-size wind turbines. As technology and engineering have advanced, the efficiency of wind turbines has grown dramatically, as has their size.

The future of wind power

Wind power is one of the lowest-cost large-scale renewable energy sources on the planet. It is also one of the largest sources of renewable energy in the United States, surpassing hydroelectric power in 2021.

The share of electricity generated from wind grew from less than 1 percent in 1990 to about 8.4 percent of all the electric power in the United States in 2020. During that year, wind power at utilities (which doesn't include agricultural and other small installations) produced 337,000 MW of electricity in the United States.

China is now the world's largest wind electricity generator. In 2019, 127 countries generated a total of about 1.42 trillion kWh of wind electricity, enough to power 1/3 of all the energy requirements of the United States. Currently, worldwide wind power generation is growing at an annual rate of about 8 percent.

⊕ **092**

Energy distribution

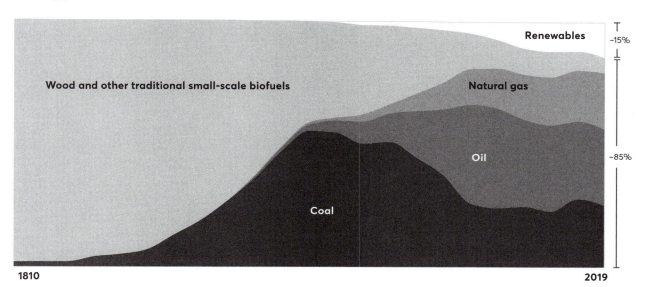

Renewables ~15%

Wood and other traditional small-scale biofuels

Natural gas

Oil ~85%

Coal

1810   2019

Solar Energy

Just ninety minutes of sunlight striking the earth's surface provides enough energy to power the world for a year. Yet solar technology at scale has not happened due to the challenges of building and installing enough solar panels to capture consistent direct sunlight.

Photovoltaic (PV) accounts for 3.1 percent of global electricity generation and is currently the least expensive renewable power source. Solar is now the third-largest renewable electricity technology behind hydropower and onshore wind.

China, the US, and India added the most new PV capacity in 2020.

How solar works

Solar panels convert sunlight into electrical energy. The sun could either strike panels directly or be reflected by mirrors to the panels. This can generate electricity directly to the grid or be stored in batteries or thermal storage.

Most solar panels are made from silicon. They work by allowing particles of light to knock electrons free from atoms, generating a flow of electricity.

Limitations and opportunities

The average house in the US uses about 11,000 kWh of electricity per year. This translates to about 30 kWh of electricity per day. A typical 60 cell solar panel is about 2 m by 1 m in size. This translates to an area of 1.7 square meters and produces about 270 to 300 W.

Assuming about five hours of peak sunlight, a panel would be able to produce about 1.3 kWh per day. In order to be able to meet the average household requirement of 30kWh, a 24 cell array would be sufficient. That would require about 40 square meters in roof/land area.

Solar energy by latitude, mid-March

Location	Value
Helsinki, Finland (60° N)	14.2
Paris, France (48° N)	15.5
San Francisco, USA (37° N)	16.4
Miami, USA (25° N)	17.2
Mumbai, India (18° N)	17.4
Bangkok, Thailand (13° N)	17.5
Accra, Ghana (4° N)	17.6
Quito, Ecuador (0°)	17.7
Lima, Peru (12° S)	17.6
Pretoria, South Africa (25° S)	17.2
Buenos Aires, Argentina (34° S)	16.8
Wellington, New Zealand (41° S)	16.1
Puerta Tora, Chile (55° S)	15.0

Measured as kiloWatts absorbed per square meter, mid-day

In 2020, about four billion megawatts of electricity was produced at electricity generation facilities in the US. About 60 percent of this power was generated by burning fossil fuels, 20 percent came from nuclear power plants, and 20 percent came from renewable energy sources, primarily wind and hydroelectric power. Solar energy supplied only 2.3 percent of the total.

Employing current panel technology, the US would require roughly 14,000,000 acres or 22,000 square miles of solar panel-filled land (roughly the size of the Mojave Desert) to have all of its energy needs satisfied by solar.

Distribution is also a consideration. Centrally located solar farms in the sunniest parts of the US would require power to be transported from where it is generated to where it is needed most. For example, a 20 gigawatt solar installation has been given planning approval in Australia's Northern Territory to supply Singapore with 15 percent of its power via an undersea cable.

Electricity can be wasted in transport. Long-distance delivery requires infrastructure and engineering to minimize this waste.

Availability and angles of sunlight

Solar energy is only generated when the sun shines. Latitude, climate, the sun's position, and weather patterns also impact the amount of solar radiation that's received. For example, the sun is lower in the sky in the winter. In a high-yield solar area such as central Colorado or the Mojave Desert, a single panel can produce 400 kWh of energy per year. In contrast, the same panel in Michigan will yield only 280 kWh. At more northerly European latitudes, yields are even lower, producing (for example) only 175 kWh annual energy yield in southern England.

ALIA-250 AIRPLANE

The Alia-250 airplane is fully electric, takes off vertically, can fly 250 miles on a charge and can be recharged in one hour.

Assuming about five hours of peak sunlight, a panel would be able to produce about 1.3 kWh per day. In order to be able to meet the average household requirement of 30kWh, a 24 cell array would be sufficient.

Solar technology advancements

As of 2021, the efficiency of solar panels varies between 11 and 24 percent. This means less than one-quarter of the available energy is converted to electricity.

Researchers at multiple labs have achieved solar conversion efficiencies as high as 50 percent. More efficient solar PV is projected to be available commercially within the next decade. If implemented, it would reduce the amount of land needed to power a community.

🌐 **091**

CARBON NEUTRAL TATTOO

How Much Land Would It Take to Power the US through Solar?

A solar farm needs four square acres to generate approximately one megawatt of energy. The US would therefore need eight million square acres—an area smaller than the Mojave Desert—in order to rely solely on solar power. That's about as much land as the country already uses for coal power generation and acquisition.

How does this compare to the land area the United States uses for other forms of power? In addition to the area used by coal, 26 million square acres are leased by oil and gas companies and 22 million more are used for corn ethanol production.

On a per-square-meter basis, solar power compares favorably with hydropower and wind power, requiring just 10 percent of the land area needed for hydropower or wind turbines to generate the same wattage.

The amount of land needed to power the entire planet with solar is about 135 million acres, which is equivalent to the size of France.

⊕ 088

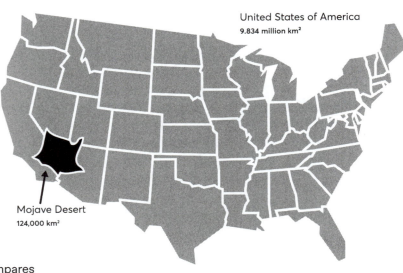

United States of America
9.834 million km²

Mojave Desert
124,000 km²

In five hours, the solar panels on these eight million acres could generate enough electricity to power the United States for the day.

Land required for generating electricity

LAND USE	ELECTRICITY GENERATED PER ACRE (GWh)		ACRES PER GWh PER YEAR[1]	RECLAMATION OPTIONS[2]
	Over 75 years	Over 25 years		
Solar	25.00	8.33	3 (perpetual)	Remove panels or dual-use
Nuclear	16.66	16.66	0.06 (only once)	Very high cost (radioactive)
Coal	11.11	11.11	0.09 (only once)	Costly (<15% reclaimed)
Wind	2.90	0.96	26 (perpetual)	Remove turbines or dual-use
Hydro	2.50	0.83	30 (perpetual)	Drain dam and restoration
Biomass	0.40	0.13	188 (perpetual)	Replant trees

[1] A GWh is the same as a million kilowatt hours.

[2] Reclamation may not be necessary because land can be simultaneously used for other purposes.

" The most important thing an individual can do is be less of an individual. Join together with other people in movements large enough to affect changes in policy and economics that might actually move the system enough to matter.

You can't do it one light bulb, one vegan dinner at a time anymore. You should do those things and do them for a whole variety of reasons like they're the morally right thing to do and they're going to save you money, and they're going to make you more healthy...But don't do them expecting that by doing them you've somehow done your duty.

What we need you to be is effective citizens moving policy.

Citizenship has not been the thing we've been best at in this country in recent years and we're paying the price in a number of places. But the most obvious probably, and the most long-term damage, is what we're doing to the physical systems of the Earth.

So that's my sense of things.

Movements are, history would indicate, the one way we have of standing up to unjust, entrenched power. "

— Bill McKibben

Advances in Solar Power

Industry innovations, scaling the number of units produced, bigger factories, and more efficient panels have all contributed to lowering photovoltaic costs.

In addition, new technologies promise significant improvements, including perovskite and bi-facial panels.

Solar power: cost per watt generated

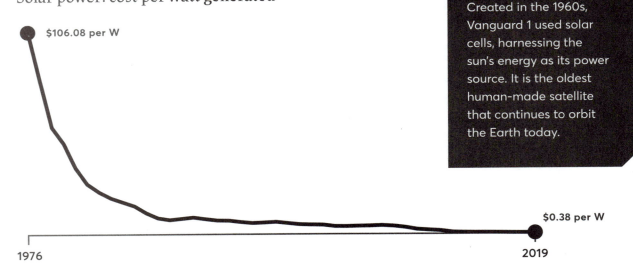

$106.08 per W

$0.38 per W

1976

2019

SOLAR SATELLITE

The Earth is continuously barraged with 173,000 terawatts of solar energy. Created in the 1960s, Vanguard 1 used solar cells, harnessing the sun's energy as its power source. It is the oldest human-made satellite that continues to orbit the Earth today.

Perovskite

Solar cells convert sun rays into electricity via semi-conductors. Currently, most cells use silicon as their semi-conducting material. Perovskite is a new material that, like silicon, is a semiconductor.

+	−
Thinner	Not as durable or reliable as silicon cells
Simpler to manufacture (can be printed)	Manufacturing contains unknown pitfalls
Better tuned to match the spectrum of the sun	Possible environmental consequences
Potentially cheaper	

Bi-facial panels

In some locations, reflected solar radiation may be available. By putting cells on both sides of the panel, energy generation can be increased by 10 percent or more.

+	−
Enables the capture of indirect radiation	Requires more investment
Increases energy captured	May require double glass
Requires less solar plant equipment	
Reduces overall costs long-term	

Widening the use of PV

Solar panels, as they become cheaper and more efficient, are being used in more locations, including floating farms, the sides of buildings, and mounted to cars.

Floating solar farms

Most of the world's solar farms are on solid ground. Developers are now experimenting with placing solar panels on floating platforms in bodies of water. This could be applied in areas where space is restricted, such as cities, and where dams or large lakes exist nearby.

Floating solar is in its early stages. 2020 saw 3 GW installed globally (compared to the global total of around 140 GW.) Challenges include getting the power generated to land, mooring platforms appropriately, and avoiding biological fouling.

Building-integrated photovoltaics

Solar farms are typically constructed away from cities and other dwellings to build large, efficient installations. As the price of panels drop, building-integrated photovoltaics (BIPV) have entered the market. BIPV utilize PV panels instead of roofing materials within windows and as part of the façade of a building.

PV panels on cars

PV cells are now being tested by directly mounting them on cars. The current efficiencies of PV panels combined with the relatively small surface area on a car means that the panels can't yet make a significant contribution to the electricity needs of the car. Prototype demo cars range from small panels driving the air conditioning on a Toyota Prius to fully-fledged "solar vehicles" like the early Tokai Challenger and the more recent Stella.

Other innovations in development include solar roadways, in which the panels are embedded directly into the road surface, and printable solar panels that are translucent and can be used as windows.

🌐 **217**

> We have many advantages in the fight against global warming, but time is not one of them. Instead of idly debating the precise extent of global warming, or the precise timeline of global warming, we need to deal with the central facts of rising temperatures, rising waters, and all the endless troubles that global warming will bring. We stand warned by serious and credible scientists across the world that time is short and the dangers are great. The most relevant question now is whether our own government is equal to the challenge.

— Senator John McCain

Hydroelectric Power

Hydropower, or hydroenergy, is currently the world's largest source of renewable electricity, supplying around 16 percent of total power—roughly three times the amount of wind power produced and six times the amount of solar. This form of renewable energy uses water stored in dams and flowing in rivers to create electricity in hydropower plants.

Compared to other renewables, hydropower is the most available, reliable, and affordable energy source. It's also among the oldest: humans have been harnessing water to perform work since the Han Dynasty in China (between 202 BC and 9 AD).

Currently, East Asia and the Pacific lead in global hydropower capacity, followed by Brazil and the United States.

Within the last decade, global electricity production from hydro has increased by two-thirds. Approximately 1,000 dams are currently under construction, mostly in Asia. According to the International Energy Agency, hydropower generation is expected to increase by another 50 percent by 2040.

The largest hydroelectric generating plant is the Three Gorges Dam in China. It has a generating capacity of 22,500 megawatts (MW), approximately the equivalent of 22 nuclear power plants, or 11 Hoover Dams. Second in size is the Itaipú hydroelectric power plant in Brazil and Paraguay.

How the motion of water is converted to electrical energy

Commercialized hydropower was developed in England in the early 19th century and has been utilized around the globe ever since. Hydropower produces electrical energy through the power of moving water. The falling water rotates the blades of a turbine, which then spins a generator, converting the mechanical energy of the spinning turbine into electrical energy. The higher the elevation and the more water flowing through the turbine, the greater the capacity for electricity generation.

Environmental impacts of hydroelectric power

Generating electric power by building dams to store water has long-term environmental impacts. Dams store water, provide renewable energy, and prevent floods. The construction of a dam also releases greenhouse gases such as carbon dioxide and methane, which form in natural aquatic systems and human-made water storage reservoirs as a result of aerobic and anaerobic decomposition of organic material.

Dams often contribute to the destruction of carbon sinks in wetlands and oceans, the deprivation of ecosystem nutrients, the destruction of habitats, and the displacement of poor communities.

⊕ **095**

The best way to predict the future is to invent it.

— Alan Kay

The world's largest dams as of 2019

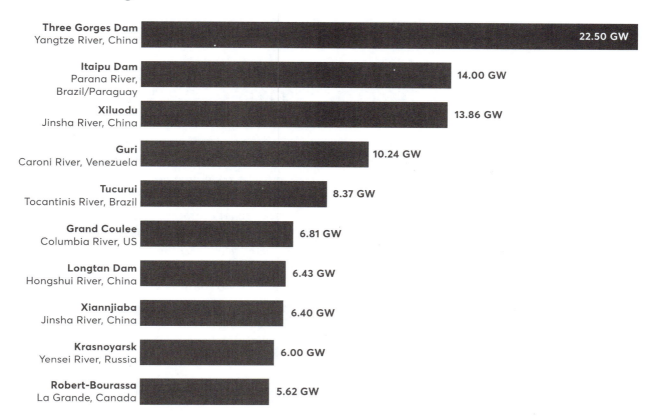

Three Gorges Dam
Yangtze River, China — 22.50 GW

Itaipu Dam
Parana River,
Brazil/Paraguay — 14.00 GW

Xiluodu
Jinsha River, China — 13.86 GW

Guri
Caroni River, Venezuela — 10.24 GW

Tucurui
Tocantinis River, Brazil — 8.37 GW

Grand Coulee
Columbia River, US — 6.81 GW

Longtan Dam
Hongshui River, China — 6.43 GW

Xiannjiaba
Jinsha River, China — 6.40 GW

Krasnoyarsk
Yensei River, Russia — 6.00 GW

Robert-Bourassa
La Grande, Canada — 5.62 GW

Newly installed capacity by region

East Asia and Pacific — 14,466 GW

Europe — 3032 GW

South and Central Asia — 1609 GW

Africa — 938 GW

North and Central America — 531 GW

South America — 476 GW

A tidal generator can produce power for up to 22 hours a day.

Generating Energy from Ocean Tides

Ocean tides are a renewable and relentless source of energy. They are powered by the gravitational pull of the moon and sun along with the rotation of the Earth.

The motion of tides causes water levels near the shore to rise and fall by as much as 12 meters between low and high tides. Using the kinetic energy of the rising and falling tides to turn the blades of an underwater turbine, electric energy can be produced with zero emissions.

Tidal power remains a tiny fraction of the world's generated electricity.

The history of converting ocean tides to energy

People have harnessed the tides and used their energy for many centuries. Tide mills are similar to water wheels. A pond fills up as the tide comes in. When the tide goes out, the pond empties, and the moving water rotates a water wheel.

Archeologists have discovered the remains of a tide mill from 619 AD at the site of the Nedrum Monastery in Northern Ireland. By the Middle Ages, tide mills were common.

In the 1700s, 76 tide mills were being used in London alone. At one point, there were 750 tide mills in operation on the shores of the Atlantic Ocean. This included about 300 in North America, 200 in the British Isles, and about 100 in France.

How modern tidal energy generators work

In a tidal generator, the retreating tides turn the propeller-like blades of a submerged tidal turbine. The blades rotate from 12 to 18 times a minute depending on tide strength. The turbine powers a

Installed electricity capacity worldwide in 2019

Source	Capacity
Fossil fuels	4213
Renewables	2497
Hydroelectricity	1140
Wind	622
Solar	584
Nuclear	369
Hydroelectric pumped storage	168
Biomass and waste	134
Geothermal	14
Tide, wave, and fuel cell	2

gearbox that turns an electric generator, which creates electricity.

An effective way to capture tidal energy is to place the turbine in a narrow channel between two landmasses. When the tide comes in, the water rises on one side of the channel and pours down the channel to the other side. When the tide goes out, the water on the higher side pours back through the channel where the water level has dropped.

Tidal power today

South Korea's Sihwa Lake Tidal Power Station is the largest in the world. With 254 MW generated every hour, it's big enough to power more than 40,000 homes. The second-largest operating tidal power plant is in La Rance, France, with 240 MW of electricity generation capacity.

Tidal power remains a tiny fraction of the world's generated electricity, but the tides themselves are widespread and persistent.

🌐 **098**

YOU CAN MAKE A DIFFERENCE

Visit **www.thecarbonalmanac.org** and sign up for **The Daily Difference**, a free email that will connect you with our community. Every day, you will join thousands of other people connecting around specific actions and issues that will add up to a significant impact.

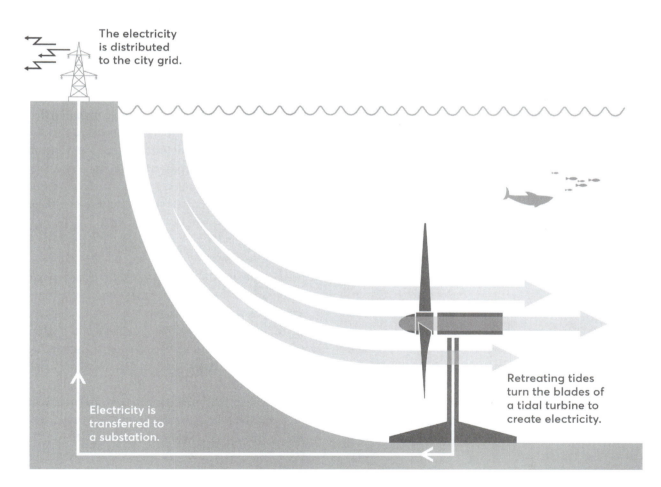

The electricity is distributed to the city grid.

Electricity is transferred to a substation.

Retreating tides turn the blades of a tidal turbine to create electricity.

Nuclear Energy by Fission

Nuclear energy by fission generates about 10 percent of the world's electricity, representing the second-largest source of low-carbon electricity after hydropower. Because nuclear power plants do not burn fuel, they do not produce greenhouse gas emissions. About 450 nuclear power reactors are in operation worldwide.

Usage of nuclear power varies from country to country. In France, for example, nearly 70 percent of the country's energy is produced by nuclear power; in Australia, that number is zero.

How fission works

Two types of physical processes produce nuclear energy from atoms: fission and fusion. Fission splits a larger atom into two or more smaller ones, whereas nuclear fusion joins two or more lighter atoms, forming a third, heavier element. Fission is the process most commonly known and referred to as nuclear power.

Nuclear fission splits uranium atoms to generate heat and produce steam, which is then used by a turbine generator to produce electricity. It is among the most effective processes to produce grid-scale quantities of electricity.

Types of fission reactors

All nuclear fission power plants make electricity from steam created by the heat of splitting atoms. But there are two different ways that steam is used.

Pressurized Water Reactors keep water under pressure so it heats but does not boil. Water from the reactor and the water that is turned into steam are in separate pipes and never mix.

In Boiling Water Reactors, the water heated by fission boils and turns into steam to turn the generator. In both types of plants, the steam is turned back into water and can be used again in the process.

How safe are nuclear reactors?

Nuclear accidents are rare, but those that have occurred—at Chernobyl, Three Mile Island, and Fukushima—have had catastrophic consequences. Research on death rates from energy production scores nuclear among the safest, just below renewables such as wind and hydropower.

Construction and fuel

Unlike fossil fuel-fired power plants, nuclear reactors do not produce air pollution or carbon dioxide while operating. However, the processes for mining and refining uranium ore and making reactor fuel do require significant amounts of carbon-producing energy.

Nuclear power plant structures and facilities also depend on substantial amounts of metal and concrete for construction, which means significant amounts of energy in the manufacturing process. This contributes a great deal of carbon because concrete is a major contributor to the planet's carbon load.

Radioactive waste

A major environmental concern related to nuclear power is the creation of radioactive wastes such as uranium mill tailings, spent (used) reactor fuel, plutonium, and other radioactive wastes. The half-lives of the radio-isotopes produced are very long—some greater than one million years. Control and management of nuclear waste poses challenges.

The most current method for nuclear waste disposal is storage, using either steel cylinders as a radioactive shield or deep and stable geologic formations. The disposal of nuclear waste by storage is controversial, as the leakage of nuclear waste may cause environmental disasters. These techniques are still under development.

The future of nuclear energy by fission

The future of nuclear power plants is undecided. The world's approximately 450 nuclear reactors are getting older. With an average usable age of 35 years, one-quarter of all nuclear power plants in developed countries will need to be shut down by 2025.

Following the Fukushima meltdown, a number of countries began to consider phasing out nuclear programs, with Germany expected to shut down its entire nuclear fleet by 2022.

The US has 95 nuclear reactors in operation, but only one new reactor has started up in the last 20 years.

However, more than 100 new nuclear reactors are being planned in other countries, and 300 more are proposed, with China, India, and Russia leading the way. The International Atomic Energy Agency estimates that nuclear energy production will double from 2019 levels of 392 billion watts to 715 billion watts in 2050.

🌐 **093**

Death rates from energy production

Coal
2462

Nuclear
7

Death rates are based on deaths from accidents and air pollution per 100 terawatt-hour (TWh).

Nuclear Energy by Fusion

Atoms create nuclear energy using two physical processes: fission and fusion. Fission is the process that produces electricity in nuclear power plants across the globe. Fusion is the energy source of the sun and stars, but it's currently very challenging to contain and manage as usable energy on Earth.

Inspired by the sun

In the 1920s, British physicist Arthur Stanley Eddington was the first to suggest that nuclear fusion powers the universe. Fusion uses hydrogen, the most abundant element in the universe, to generate energy in the form of heat and light. Its waste products are helium and tritium.

Fusion is clean, safe, powerful, and efficient and has been in constant use for about 4 billion years by the sun. The Earth has enough of fusion's primary fuels—heavy hydrogen and lithium, which are found in seawater—to last 30 million years. If fusion could be harnessed effectively, it could potentially generate endless amounts of pure and waste-free energy.

How fusion works

Unlike fission, which creates energy by splitting atoms apart, fusion occurs when two atoms combine and fuse to form a third, heavier element. In the process, some of the atoms' mass is converted into a massive output of energy.

Challenges of nuclear fusion

Earth does not have the same intensity of gravity as the sun, so a fusion reactor needs a different way to contain the reaction. A practical fusion reactor would require ten times more heat than found on the sun.

To produce a fusion reaction on Earth, the intense heat and pressure found in the sun's core needs to be reproduced. This intense heat, as well as internal pressures needed to fuse the atoms, requires sufficient containment to hold and maintain that fusion reaction long enough for a net power gain. As no material is capable of containing such a hot and dense plasma where the fusion reaction is taking place, magnets far more potent than those used in an MRI machine suspend the plasma in a special doughnut-shaped apparatus known as a Tokamak. Unfortunately, a Tokamak is not very efficient at maximizing energy output, and research into more advanced magnetic confinement geometries is a significant area of interest for nuclear physicists.

Ongoing fusion research and development

More than 50 countries are researching nuclear fusion and plasma physics. Fusion reactions have been successfully achieved in many experiments, though none as yet has demonstrated a net power gain.

In southern France, 35 nations are collaborating on project ITER ("The Way" in Latin) to build the world's

Useful fusion power has been ten years away for the last sixty years.

— Unknown physicist

largest Tokamak to prove the feasibility of fusion as a large-scale and carbon-free energy source.

ITER holds the possibility of being the first fusion device to produce more energy output than is required to run the reactor and currently holds many records for sustained reactions.

In China, the China Fusion Engineering Test Reactor (CFETR) will be slightly larger than ITER, with a radius of about seven meters. As of 2020, CFETR was in the design and technology prototyping phase with construction to begin later in the decade. Initially, it will demonstrate fusion operation at about 200 MW fusion power and will eventually upgrade to at least 2000 MW of fusion power and 700 MW of net output.

By 2021, investors began embracing the possibility of fusion energy as never before. Here are some firms and the money they've raised.

⊕ 094

Fusion funding

NAME	FUNDING (US $)	APPROACH	LOCATION
Avalanche Energy	33 Million	Palm-sized reactors	Seattle, WA USA
Commonwealth Fusion Systems	2.5 Billion	Tokamak	Cambridge, MA USA
ENN Energy Research Institute	Public Company	Field Reversed Configuration	Langfang, Hebei China
First Light Fusion	25 Million	Impact Inertial Confinement	Yarnton UK
General Fusion	322 Million	Liquid Liner Compressor	Burnaby, BC Canada
HB11	4.8 Million	Laser Boron Fusion	Sydney, NSW Australia
Helion Energy	2.7 Billion	Field Reversed Configuration	Redmond, WA USA
Lockheed Martin Skunk Works	Public Company	Compact (truck-sized) reactors	California USA
TAE Technologies	90 Million	Field Reversed Configuration	Foothill Ranch, CA USA
Tokamak Energy	10 Million	Spherical Tokamak	Abingdon UK
Zap Energy Inc.	50 Million	Z Pinch	Seattle, WA USA

CARBON IS EVERYWHERE

Some of the common items in our homes made from carbon include:

Asphalt	Electric blankets	Lipstick	Soft contacts
Aspirin	Eye glasses	Pajamas	Solar panels
Bandages	Fertilizers	Pillow filling	Tents
Cell phones	Food preservatives	Shampoo	Toothpaste
Deodorant	Hand lotion	Shaving cream	Toys
Detergent	Heart valves	Shoes/sandals	Vitamin capsules

Geothermal Energy

The Earth's core is hot. By tapping that heat, engineers can put it to use.

Water and/or steam carries the geothermal energy to the earth's surface where it can be used for heating and cooling or harnessed to generate clean, renewable electricity.

Globally, geothermal energy provides significant levels of electricity in countries such as Iceland, El Salvador, New Zealand, Kenya, and the Philippines. More than 25 percent of Iceland's electricity is generated this way, and most of its heating as well.

The United States leads the world in the production of geothermal electricity, producing more than 3.5 billion watts—enough to power about 3.5 million homes. Indonesia, the Philippines, Turkey, and New Zealand complete the top five geothermal producers.

How geothermal electricity works

Geothermal-based power plants use steam produced from geothermal reservoirs to generate electricity. Three geothermal power plant technologies are currently used to convert hydrothermal fluids to electricity:

- dry steam
- flash steam
- binary cycle

Advantages and disadvantages of geothermal energy

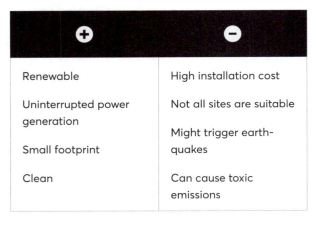

⊕	⊖
Renewable	High installation cost
Uninterrupted power generation	Not all sites are suitable
Small footprint	Might trigger earthquakes
Clean	Can cause toxic emissions

The world's largest geothermal power plant complex is located north of San Francisco. It contains 22 power plants powered by steam from 350 wells and produces about 900 megawatts of electricity—more than enough to power 725,000 homes.

Geothermal power plants are compact, using 1 to 8 acres per MW, as compared to 5 to 10 acres for nuclear power plants and 19 acres for coal power plants (not including the coal mines themselves).

Environmental impacts of geothermal energy production

Most geothermal facilities use *closed-loop* geothermal systems that produce minimal emissions. This process pumps extracted water directly back into the geothermal reservoir after it has been used for heat or electricity production.

Open-loop geothermal power plants release steam and thermal effluents into the environment and can impact water and air quality. Hot water pumped from underground reservoirs often contains high levels of sulfur, salt, and other minerals. Open-loop geothermal systems emit hydrogen sulfide, a gas that has a distinctive "rotten egg" smell.

Emitted hydrogen sulfide converts into sulfur dioxide (SO_2). Small acidic particulates are created, which can cause heart and lung disease. Acid rain is caused by SO_2, and it damages crops and also acidifies lakes and streams.

Though the side effects are noteworthy, SO_2 emissions from geothermal plants are about 30 times lower per megawatt-hour in the United States than those created by coal plants—the largest SO_2 source.

Because geothermal plants remove water and steam from reservoirs within the earth, the land above those reservoirs can sometimes sink slowly over time and cause surface instability. Most geothermal plants re-inject used water into the earth via an injection well to reduce that risk.

Installed geothermal capacity, 2020

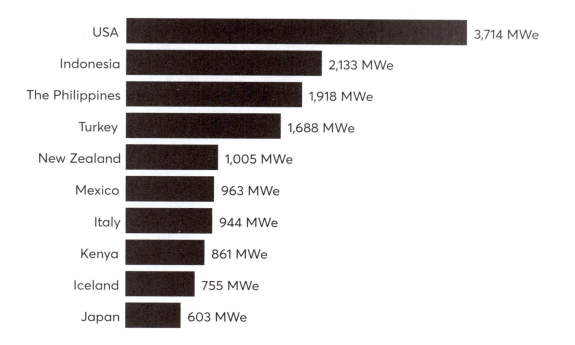

USA	3,714 MWe
Indonesia	2,133 MWe
The Philippines	1,918 MWe
Turkey	1,688 MWe
New Zealand	1,005 MWe
Mexico	963 MWe
Italy	944 MWe
Kenya	861 MWe
Iceland	755 MWe
Japan	603 MWe

Increased earthquakes are also a risk during geothermal operations. Geothermal power plants are typically located near fault zones or geological "hot spots" that are especially prone to instability. Drilling deep into the earth and removing water and steam can sometimes trigger these disturbances.

Enhanced geothermal systems (EGS)
In areas where there is sufficient heat but no natural permeability, engineers are experimenting with EGS. Fluid is injected into the subsurface under high pressure, which causes pre-existing fractures to re-open, creating permeability. This allows fluid to circulate throughout the now-fractured rock and to transport heat to the surface where electricity can be generated.

⊕ **097**

The worthwhile problems are the ones you can really solve or help solve, the ones you can really contribute something to...
No problem is too small or too trivial if we can really do something about it.

— Richard Feynman

Hydrogen for Energy Storage

Solar and wind power are now becoming cheaper than coal. But unlike coal, their power can't be stored for long and so are mostly used on the spot. As a result, renewable energy is ill-suited for changes in energy demand.

Luckily, hydrogen is an energy-storing option for renewable energies. In fact, energy can be stored in carbon-free hydrogen for long periods and then used when needed. This stored hydrogen provides readily available energy for changes in energy demands like the cold winter season when demand peaks.

It's possible to increase the amount of hydrogen storage without building more infrastructure. If hydrogen is stored in pipelines, adding pressure to the pipes can increase their storage capacity.

Hydrogen is seen by many as key for the transition to renewable energies due to their symbiotic relationship. Renewable energy can be used to produce hydrogen, which can then be used to store the energy produced.

⊕ **238**

"The four-bladed ones are especially good luck."

Energy from Biomass and Trash

Garbage can be burned to create power. This is called Waste To Energy or Municipal Solid Waste (MSW).

In 2019, the United States burned 31 million metric tons of MSW to generate electricity. This created nearly 13 billion kilowatt hours of electricity, enough to power more than a million homes for an entire year.

The method is simple: garbage is burned in an incinerator, generating heat. The heat boils water and the steam drives a generator.

The most common fuels are biomass or biogenic (plant or animal) materials. These include:

- paper and cardboard
- food waste
- grass clippings
- leaves
- wood
- leather products

In addition, some incinerators are built to handle non-biomass combustible materials like plastics and other synthetic materials made from petroleum.

Biomass energy

Humans have always burned wood for heat and cooking. Today the process is more sophisticated, using many materials that come from plants and animals. Some incinerators can even burn the fumes from landfills to create power.

The biomass used this way is carbon-based, and often contains hydrogen, oxygen, nitrogen, and small amounts of other atoms. The carbon in the biomass came from plants sequestering CO_2, and by burning it, that CO_2 is released back into the atmosphere.

Techniques for converting biomass to energy

- **Direct combustion** burns biomass to produce heat. It's the most prevalent approach.
- **Thermochemical conversion** of biomass uses pyrolysis and gasification to produce solid, gaseous, and liquid fuels. More complex than direct combustion, it involves heating biomass in pressurized sealed containers at high temperatures.
- A **chemical conversion** approach called transesterification converts vegetable oils, animal fats, and greases into fatty acid methyl esters, which are then used to produce biodiesel.
- **Biological conversion** can create ethanol, biogas, or biomethane. It involves using anaerobic digesters at sewage treatment plants and at dairy and livestock operations. In some cases, the conversion can happen at solid waste landfills. Once created, the final product is a direct substitute for traditional fuels.

Is biomass energy carbon neutral?

Biomass releases carbon dioxide to the air, but it relies on recently captured carbon from the fast cycle. Fossil fuels and other sources of power, on the other hand, release carbon that has been stockpiled for millions of years.

> **In order to be useful in dealing with climate change, any biomass process also needs to invest in creating new plants to take the place of the biomass that was turned to fuel.**

Old-growth forests, for example, are efficient stores of carbon, but can take decades or more to replace. Once cut down, rainforests are much more difficult to replace, and there can be significant issues with flooding, erosion, and other degradations to the environment. And incinerating plastic can't be neutral because the plastic that's being burned was made from fossil fuels.

⊕ **096**

Carbon-Neutral Fuels: Ammonia

Carbon-neutral fuel produces no net (additional) greenhouse gas emissions or carbon footprint. By converting carbon dioxide into a useful fuel, the carbon cycle continues without digging any new carbon out of the ground.

Carbon-neutral fuels can broadly be grouped by:
· **synthetic fuels:** made by chemically hydrogenating carbon dioxide
· **biofuels:** produced using natural CO_2-consuming processes like plant photosynthesis

The carbon dioxide used to make synthetic fuels may be directly captured from the air, recycled from power plant flue exhaust gas (combusted in industrial furnaces), or derived from carbonic acid in seawater. Common examples of synthetic fuels include hydrogen, ammonia, and methane.

Ammonia is a molecule composed of one nitrogen atom and three hydrogen atoms in a trigonal pyramid shape. It is a colorless gas with a distinct pungent smell.

In nature, it is a standard waste product of nitrogen respiration, particularly among aquatic organisms.

> **Ammonia contributes significantly to the planet's nutritional needs by serving as a precursor to 45 percent of the world's food and fertilizers.**

Ammonia, either directly or indirectly, is a building block for the synthesis of many pharmaceutical and cleaning products, including window and glass cleaners, all-purpose cleaners, oven cleaners, and toilet bowl cleaners. It can act as a refrigerant gas to be used in the purification of water, and in the manufacture of plastics, explosives, textiles, pesticides, and dyes.

By volume, ammonia carries 70 percent more energy than liquid hydrogen and nearly three times as much

PRODUCING AMMONIA

Traditionally produced ammonia generates vast amounts of carbon dioxide. Emissions can be reduced during the production process by substituting a renewable energy source such as wind or solar for fossil energy. The process uses carbon capture and storage to separate and sequester most of the CO_2 emissions created.

By weight, ammonia carries more than 20 times as much energy as today's lithium batteries.

> We need to continue holding leaders accountable for their actions. We cannot keep quiet about climate injustice. Your actions matter. No action is too small to make a difference and no voice is too small to make a difference. Let us keep the faith for the future.
>
> — Vanessa Nakate

energy as compressed hydrogen gas. By weight, it carries more than 20 times as much energy as today's lithium batteries.

Liquid ammonia can be stored in large tanks at room temperature and is safer than propane and as safe as gasoline. More than 200 million tons of ammonia are produced each year and distributed globally via pipelines, tankers, and trucks.

The use of ammonia as a renewable fuel is not without some of the drawbacks endemic to high-temperature combustion processes like the modern diesel engine. Chief among them is the potential for NO_x formation in nitrogen-rich oxidizing environments requiring extra steps in the exhaust process to reduce levels of nitrous oxide.

🌐 **109**

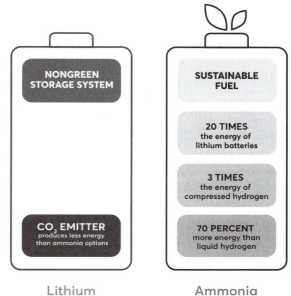

NONGREEN STORAGE SYSTEM

CO₂ EMITTER
produces less energy than ammonia options

SUSTAINABLE FUEL

20 TIMES
the energy of lithium batteries

3 TIMES
the energy of compressed hydrogen

70 PERCENT
more energy than liquid hydrogen

Lithium battery   Ammonia fuel cell

AMMONIA AS A FUEL

Ammonia can be used as fuel in three ways:
- burned directly in an internal combustion engine with zero net carbon emission.
- converted to electricity directly in an alkaline fuel cell.
- cracked to provide hydrogen for non-alkaline fuel cells.

Carbon-Neutral Fuels: Hydrogen

Fuel is used to release energy where and when it's needed, whether to cook a meal or to fill the gas tank. Carbon-based fuels take stored carbon from the slow cycle (in the form of coal or oil or natural gas) and release carbon dioxide when burned.

Carbon-neutral fuel, by contrast, produces no net (or additional) greenhouse gas emissions or carbon footprint. Instead of unlocking stored carbon, they rely on another method of releasing energy.

Carbon-neutral fuels can broadly be grouped as:
- Synthetic fuels made by chemically hydrogenating carbon dioxide.
- Biofuels produced using natural CO_2-consuming processes like photosynthesis.

In both cases, carbon dioxide is captured and then released and the net result is no change in the amount of carbon produced. The energy is stored and used where needed.

The carbon dioxide used to make synthetic fuels is captured directly from the air, recycled from power plant flue exhaust gas (fossil fuels combusted in an industrial furnace), or derived from carbonic acid in seawater. Common examples of synthetic fuels include hydrogen, ammonia, and methane.

Hydrogen as a fuel

Hydrogen can be produced from diverse domestic (or in-country) resources with the potential for near-zero greenhouse gas emissions. Once produced, hydrogen generates electrical power in a fuel cell, emitting only water vapor and warm air.

Safely and efficiently burning hydrogen is not a stumbling block. Efficiently creating, storing, and distributing it is.

Producing hydrogen

To produce hydrogen on Earth, decoupling its atoms from other materials—water, plants, or fossil fuels must occur. How this decoupling is done determines hydrogen's sustainability.

Today most hydrogen is produced from fossil gas in a pollution-heavy process using a catalyst to react methane and high-temperature steam. This process,

> ### HYDROGEN
>
> Is the most common element in the universe. Approximately 90 percent of all atoms are hydrogen. On Earth, hydrogen is rarely found in its pure form, and something else has to be processed to extract the hydrogen.

Action on behalf of life transforms. Because the relationship between self and the world is reciprocal, it is not a question of first getting enlightened or saved and then acting. As we work to heal the earth, the earth heals us.

— Robin Wall Kimmerer

known as steam methane reforming, results in hydrogen plus carbon monoxide and a small amount of carbon dioxide, and is used to refine oil and produce fertilizer.

Hydrogen can also be produced through the electrolysis of water, leaving nothing but oxygen as a byproduct. Electrolysis employs an electric current to split water into hydrogen and oxygen inside an electrolyzer. If renewable energy such as solar or wind produces the electricity, the resulting pollutant-free hydrogen is called *green hydrogen*.

Fuel cell electric vehicles

Liquid or highly-compressed hydrogen powers fuel cell electric vehicles (FCEVs). These vehicles are two to three times more efficient than conventional internal combustion engine vehicles and produce no tailpipe emissions other than water vapor.

Fuel storage

The energy in 2.2 pounds (1 kilogram) of hydrogen gas is about the same power in one gallon (6.2 pounds, 2.8 kilograms) of gasoline. Because hydrogen has a low volumetric energy density, it is stored onboard a vehicle as a compressed gas or liquid to achieve the driving range of conventional vehicles. Storing hydrogen compactly is a challenge because doing so requires high pressures, low temperatures, or chemical processes.

🌐 110

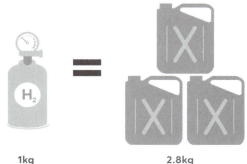

1kg   2.8kg

LED ENERGY SAVINGS

LED lighting uses 75 percent less energy and lasts up to 25 times longer than incandescent lighting. By 2035, annual energy savings from LED will equal the annual energy output of 1,000 power plants.

Job Transition from Fossil Fuels

In 1859, they struck oil in Titusville, Pennsylvania.

Exploring for oil in places where skeptics are doubtful is called wildcatting. With luck, a wildcatter becomes a millionaire overnight.

Once oil is struck, extracting it is a difficult, dirty but steady job. The wildcatters earn huge profits and often pay the people who work in the fields handsomely.

The extraction economy has produced millions of well-paying jobs, and entire communities and cultures have been built around the work of pumping oil and mining coal.

Renewable energy is efficient and clean, but it doesn't have similar economic or cultural dynamics. No one is getting rich overnight by finding a place where the sun shines or the wind blows. And while it's true that jobs installing and maintaining these plants are healthier and more resilient, they represent a significant cultural shift. Mineworkers don't always eagerly sign up for retraining in new industries.

Concern about job loss in the fossil fuel industry is based on the sudden transition of established industries. At the same time, accelerating the transition to renewable energy will create job opportunities in the installation and maintenance of energy generation and storage technologies, energy efficiency enhancement and improvements to the energy grid.

---

25 percent of all the fossil fuel hubs within the US are also ideal sites for renewable energy production.

Many communities are built around fossil fuel extraction, not only providing high-paying jobs to those with lower formal education requirements but also tax revenues and downstream jobs. However, 25 percent of all the fossil fuel hubs within the US are also ideal sites for renewable energy production. Providing financial assistance and training to people in these regions has been proposed to assist in ensuring a smooth transition to a renewables-led future. The economic cost of such a program is expected to be small when compared with other social security schemes.

In 2020 the US Bureau of Labor Statistics counted about 43,000 coal miners (down from 84,000 in 2010), while in 2020, 90 percent of energy generation capacity added to the grid was renewable in nature.

🌐 **070**

Fossil fuel employment %

| 0 | 20 | 40 | 60 | 80 | 100 |

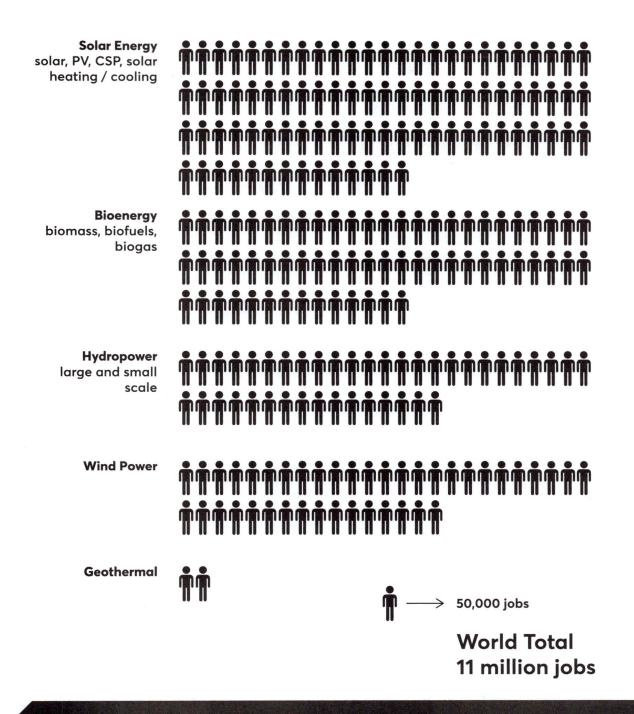

Solar Energy
solar, PV, CSP, solar
heating / cooling

Bioenergy
biomass, biofuels,
biogas

Hydropower
large and small
scale

Wind Power

Geothermal

→ 50,000 jobs

**World Total
11 million jobs**

AN ICONIC BUILDING GOES RENEWABLE

The Empire State Building (and others in the same realty group) are now powered completely by wind. Switching these buildings to clean electricity will save 200,000 metric tons of CO_2—the equivalent of removing all the taxis in New York City from the road for a year.

The Challenges of Critical Mineral Needs for Clean Energy

As we move away from fossil fuels to renewable sources of power, there will be a shift in the minerals needed to power the planet.

Minerals are needed for:
- electric vehicles (EV)
- energy storage
- solar photovoltaic panels
- wind farms

An electric car uses six times the minerals used by a combustion engine car. And a land-based wind farm uses nine times the minerals used by an electrical plant that is gas-fueled.

Batteries in the new technologies require lithium, cobalt, nickel, manganese and graphite. The permanent magnets in EV motors and wind turbines need rare earth elements. And copper and aluminum are necessary for electrical networks.

The demand for these minerals is forecast to increase significantly. Demand for lithium is expected to increase as much as 90 percent in the next two decades. Nickel and cobalt are expected to increase 60-70 percent, and copper and rare earth elements over 40 percent. In the case of lithium alone, the market for lithium-ion batteries is expected to increase from its 2017 high of $30 billion to $100 billion by 2025.

Currently, production levels are set to double for these minerals by 2040, while achieving net zero by 2050 would require six times today's production levels. Based on current supply and future investment and production plans, the transition to a renewable energy system faces risks in terms of timelines and financial projections.

Challenges and vulnerabilities include:
- geographic concentration of mineral resources
- concentration in ownership and control of mines
- complexity of the supply chain
- long lead-times for project development and ramp-up at mines
- decline in ore quality causing more energy, expense and emissions for extraction
- growing demand to source sustainable and responsibly-produced minerals
- increasing climate risks (water stress, heat stress, and flooding) to mining operations

For example, most of the challenges and vulnerabilities apply in the case of the largest producer of cobalt in the world, the Democratic Republic of the Congo (DRC). The battery in some Teslas depends on 10 pounds of cobalt (400 times the cobalt in a cell phone). The human rights and environmental issues at its mines currently threaten its strategic position in the new renewable energy economy and threaten its citizens as well.

Ford is now spending billions to develop US plants and processes to use lithium iron phosphate as a cobalt substitute. And the DRC is working to take control of the issues in its own country.

We have an amazing potential for renewable energy, be it through our strategic metals or through our rivers. How can we put this amazing resource at the disposal of the world, but while making sure that it first benefits Congolese and it benefits Africans?

— Félix Tshisekedi, President of the Democratic Republic of the Congo

Across the western United States, tribal nations are experiencing the challenges of new renewable energy systems. Extraction of critical minerals is often harmful to humans and the environment. Long-standing mining laws in the United States have traditionally worked against the tribal communities and their lands despite treaty rights to hunt, fish, and forage there.

The mining challenges facing Indigenous people in the US include:

- open-pit gold mines producing antimony for future use in solar-powered electrical grids
- open-pit and underground copper mines for standard electrical grids
- mining projects for lithium for EV batteries

These and other activities threaten the lands of the Nez Perce, the Tohono O'odham, the Pascua Yaqui, the Hopi, the Fort McDermitt Paiute and Shoshone, the Hualapai, and the San Carlos Apache.

⊕ **264**

The taste of "success" in our world gone mad is measured in dollars and francs and rupees and yen. Our desire to consume any and everything of perceivable value—to extract every precious stone, every ounce of metal, every drop of oil, every tuna in the ocean, every rhinoceros in the bush—knows no bounds. We live in a world dominated by greed.

We have allowed the interests of capital to outweigh the interests of human beings and our Earth... It makes no sense to invest in companies that undermine our future. To serve as custodians of creation is not an empty title; it requires that we act, and with all the urgency this dire situation demands.

— Archbishop Desmond Tutu

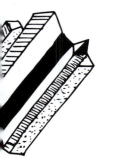

COAL TO SOLAR

The Martiki solar plant is built on the site of a former coal mine in Kentucky. It will power 33,000 homes and is being built by former coal miners.

Surprising Facts About Meat

SERVING SIZE

The British Dietetic Association's suggested portion of meat is three ounces—the size of a deck of cards.

LAND USE

The livestock production process, which requires grazing pastures and feed growth, accounts for almost 80 percent of agricultural land use.

SAVINGS WITHIN REACH

If everyone on the planet became a vegetarian, we could save an estimated $1.6 trillion in CO_2 and health damages by the year 2050. If we all went vegan, that number could jump to nearly $1.8 trillion.

WATER FOR BEEF

Raising beef for consumption requires 15,415 liters of water per kg—nearly 48 times as much as the average amount needed to grow vegetables.

ACTUAL CONSUMPTION

In 2020, the average American ate almost 12 ounces daily—*four* decks' worth of meat.

CHEAP MEAT

The US meat industry gets $38 billion a year in federal subsidies (by comparison the fruit and vegetable industry receives only $17 million annually). That's why McDonald's can sell $2 cheeseburgers. Worldwide, subsidies are over $500 billion and are expected to grow to more than a trillion.

⊕ 243

Leave the world better than you found it, take no more than you need, try not to harm life or the environment, make amends if you do.

— Paul Hawken

Food Loss and Waste

Every year, more than one billion tons of food is lost or wasted. Food can be *lost* in the early stages of the food value chain, during production, storage, processing, or distribution. Food is considered *wasted* if once edible it is not eaten.

One estimate by the Food and Agriculture Organization of the United Nations (FAO) puts the **food lost and wasted at a third of all food produced.** The FAO estimates that 4.4 gigatons of CO_2, or about 8 percent of total anthropogenic (human-caused) greenhouse gas (GHG) emissions, are emitted annually due to food waste and loss. If food waste were a country, it would be the third-largest emitter in the world.

Every time food is wasted, so is the energy that went into producing it. If the food decomposes in a landfill, it also contributes to harmful methane levels.

Contribution to greenhouse gas emissions

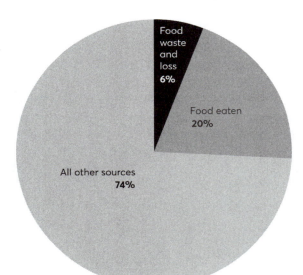

Food waste and loss 6%

Food eaten 20%

All other sources **74%**

Here are six steps people can take to reduce the impacts of food waste:

1. Plan ahead to purchase foods that have a lower carbon footprint and plan to reduce waste. Check cupboards before grocery shopping and organize meals with already-stocked ingredients, and build menus that use leftovers.
2. Prepare meals with shelf life in mind. If something won't keep well, make only what can be consumed.
3. Portion size can lead to wasted food and money. Reduce the portion size if making meals too big to consume.
4. Preserve foods by pickling, stewing, or freezing foods.
5. Prioritize your fridge or pantry often and reduce the likelihood of waste by using the oldest items first. Then use curbside or backyard composting for anything that does spoil to reduce the harmful methane that is created when food rots in landfills.
6. Promote food preservation on social media or with family and friends. Share favorite food-saving recipes and techniques to reduce waste and decrease climate impacts of food consumption.

⊕ **031**

> The major problems in the world are the result of the difference between how nature works and the way people think.
>
> — Gregory Bateson

ABOUT THIS ICON

The Almanac is based on thousands of sources. Don't take our word for it. Look for this number at the end of an article and then visit www.thecarbonalmanac.org/999 (but replace 999 with your article number). **Dig deep and share what you learn.**

Using Agriculture as a Carbon Sink

Since the dawn of agriculture, farming is estimated to have released 133 gigatons of carbon from the soil.

Regenerative agriculture is said to grow more nutrient-dense food while increasing productivity and profitability for farmers. Using a mix of ancient farming techniques and modern technology, regenerative agriculture has the potential to transform portions of the food production system into a carbon sink, thereby absorbing CO_2. This agricultural method reinforces food systems that are more resilient to climate instability as well as restores carbon content through:

· **Carbon farming:** Crop rotations, diverse cover crops, and zero or minimal fertilizers.
· **Carbon ranching:** High-intensity planned rotational grazing of livestock to reclaim grasslands. Manure nurtures the soil and the microorganisms that live there.

At its current rate of growth, by 2050 regenerative agriculture is estimated to reduce the amount of CO_2 by 23.15 gigatons and is ranked one of the most impactful solutions for climate change.

⊕ **218**

NYC
Skyline

59.4 KM³

How much is
133 gigatons of
carbon dioxide?

133 GT of CO_2 weighs 1,330 times more than all of Manhattan. It's a cube of coal 59.4 km on a side—2.72 times the length of the borough and 136.57 times taller than the highest building.

CO_2 (ppm) by year

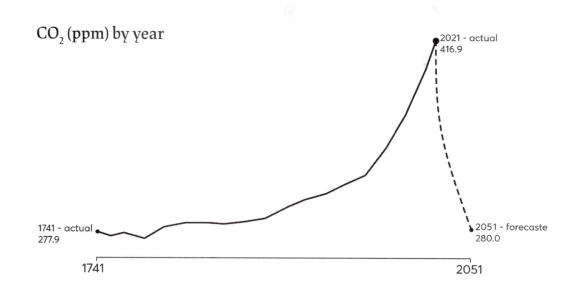

2021 - actual
416.9

1741 - actual
277.9

2051 - forecaste
280.0

1741

2051

Developing Plant-Based Alternatives

If all humans followed the diet of the United States, every hectare of the world's habitable land would need to be converted to agriculture and we would still be 38 percent short of the land needed to feed everyone.

Livestock takes up nearly 80 percent of global agricultural land, yet produces less than 20 percent of the world's supply of calories. Alternatively, if everyone adopted the plant-based diets consumed across South Asia, Sub-Saharan Africa, and some Latin American countries, the newly available landmass could feed the entire planet.

Countries with diets high in meat and animal-related food commodities require more space for livestock and animal feed production. Countries with diets centered on vegetables, grains, and seafood take up the least amount of land mass.

⊕ **099**

Land use per 100g of protein

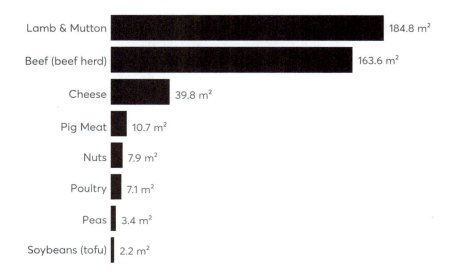

Lamb & Mutton — 184.8 m²
Beef (beef herd) — 163.6 m²
Cheese — 39.8 m²
Pig Meat — 10.7 m²
Nuts — 7.9 m²
Poultry — 7.1 m²
Peas — 3.4 m²
Soybeans (tofu) — 2.2 m²

Drip Irrigation

In 1930, a young boy in the Middle East discovered that one tree in a row had grown much taller than the rest, thanks to an aberration: a tiny pipe was leaking underground and watering the roots of this particular tree.

Thirty years later, Israeli scientists pioneered modern drip irrigation at scale. A combination of arid climate and water scarcity led to its widespread adoption both there and elsewhere.

Drip irrigation slowly trickles water directly onto roots, minimizing evaporation compared to the open canals that are traditionally used. Compared to conventional irrigation methods, by bringing just the right amount of water to the right spot, drip irrigation reduces water requirements by about 60 percent and increases crop yield by about 90 percent.

However, there are hurdles:
- standard methods require investment to set up—an estimated $2,000-$4,000 per acre, plus energy costs
- underground pipes need to be monitored for blockages
- adopting new techniques can be a challenge, particularly for small farms

In 2015, US farmers used 118 billion gallons of water a day for irrigation. But less than 9 percent of the acres irrigated used drip irrigation. Globally, that percentage drops to less than 5 percent.

The latest innovations in agriculture include:
- **Internet of Things (IoT) sensors:** Smart-tech humidity monitors embedded in the soil activate irrigation systems when the moisture content drops below a certain threshold.
- **N-Drip:** Invented by Israeli soil physics professor Uri Shani in 2017, N-Drip technology doesn't require outside energy sources or complicated machinery. It relies instead on gravitational pull to create the pressure needed to irrigate. This means it uses less water and costs a fraction of standard methods—a potential boon for small farmers. N-Drip technology is currently operational in 17 countries, including the US, Australia, Vietnam, and Nigeria.

⊕ **248**

Does Farm Size Matter?

The Earth is home to more than 570 million farms of varying sizes. In China's last agricultural census, 93 percent of the country's 200 million farms were a hectare (about 2.5 acres) or less. In the US, a "small farm" is considered by some to be "just a hundred acres," far larger than a small family farm in most other countries. While the US, Brazil, and the UK lead in farm size, large populations in India, Ethiopia, and Vietnam feed themselves with farms of a hectare or less.

> **SMALL FARMS TEND TO USE**
> · Fewer machines
> · Greater variety of crops per acre
> · Manure or compost as fertilizers, instead of relying on manufactured supplements

Despite this distribution, 88 percent of the world's farmland is on large farms of 2 hectares or more.

Industrial farming seeks efficiency and profit per acre, which is why large farms predominate in countries that have seen a rise in this approach to farming. Consolidation and scale improve short-term profits.

Very small farms are generally run by families to feed themselves. Any surplus is used to create income for the family.

Small farms (under 2 hectares) generally produce more yield per acre than large ones in similar locales. A farmer with less land is more likely to spend time and energy on every available plot. However, industrial farms in places like the US can produce as much as 10 times per acre as small farms in developing countries like India.

While the short-term yield of an industrial farm is higher, the small farm generates a smaller carbon impact.

A substantial meta-study that collated the results of more than 350 research projects has shown that no-till farming sequesters significantly more carbon than the intensive industrial farming that characterizes a large farm.

⊕ **215**

Average farm size across the globe, as of 2000

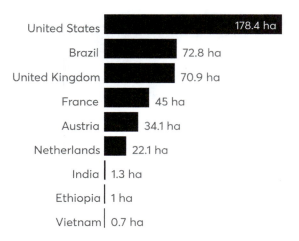

United States	178.4 ha
Brazil	72.8 ha
United Kingdom	70.9 ha
France	45 ha
Austria	34.1 ha
Netherlands	22.1 ha
India	1.3 ha
Ethiopia	1 ha
Vietnam	0.7 ha

But in all my experience, I have never been in any accident... or any sort worth speaking about. I have seen but one vessel in distress in all my years at sea. I never saw a wreck and never have been wrecked nor was I ever in any predicament that threatened to end in disaster of any sort.

— E.J. Smith, Captain, RMS Titanic

Peanut butter requires less than a quarter of the greenhouse gases needed to make a similar amount of cheese.

Chocolate and Climate

In addition to being delicious, chocolate is grown on long-lived trees near the equator, making it a seemingly perfect crop for addressing climate challenges. However, a number of factors complicate the future of chocolate.

Some of the poorest farmers in the world grow chocolate, and generational shifts and economics are driving many of them to switch to palm oil and other less sustainable types of farming. At the same time, large corporations like Nestle have industrialized production, driving poor farmers to develop monocrops and sell their beans at commodity prices. There are large-scale and persistent issues of child labor associated with this commodity pricing as well.

Increased heat threatens cacao growing regions like the Ivory Coast and Ghana which produce as much as a third of the world's chocolate. As heat rises (if humidity remains stable), transpiration causes plants to release more moisture, which can harm the trees and their yield.

In one study that examined 294 traditional chocolate-growing sites, only 31 showed increasing stability for cacao production. The others would need to be moved, altered, or face decreased production.

But there are factors that may work in favor of the chocolate industry too. A traditionally-planted hectare of cacao trees stores more than 200 metric tons of carbon. Cacao trees can, with care, last for generations, extending the cycle of carbon sequestration.

Moreover, a growing number of small producers like Original Beans have committed to eliminating child labor, doubling wages, and preserving the environment with more sustainable techniques, such as the Brazilian system called *cabruca*. Using the *cabruca* technique, cacao farmers leave most of the original rainforest trees and place the smaller cacao trees underneath the remaining canopy, providing them with cooling shade and also capturing more carbon. One study found that *cabruca* can more than double the amount of carbon captured by land that includes cacao trees.

The problem, however, is yield. Due to the lower profit per tree of a plantation that uses the *cabruca* technique, the cacao is more expensive, which requires consumers to pay more for their chocolate, or for the farmers to receive compensation in some other way in order to compete with higher yield farms. Carbon credits could be a solution to save humanity's sweetest pleasure.

🌐 **233**

About Milk and Its Alternatives

Milk has been a central ingredient of peoples' diets since humans coexisted with mammoths. Whereas most mammals lose their ability to digest traditional milk after infancy, most humans can drink it as adults.

> ### QUIK VS. 747
>
> Milk is responsible for more greenhouse gas emissions than all airplanes combined.

Traditional milk comes from cows, which produce methane throughout their lives. Methane is a powerful greenhouse gas.

In the last few decades, new alternatives to animal milk have been developed. Plant-based milk is often fortified with vitamin B_{12}, calcium, and other nutrients. Traditional milk is often fortified with additives too, like vitamins A and D.

⊕ **236**

Milk alternatives

Environmental impact

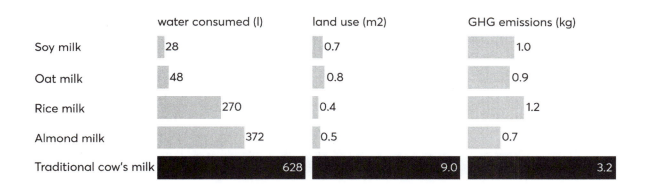

	water consumed (l)	land use (m2)	GHG emissions (kg)
Soy milk	28	0.7	1.0
Oat milk	48	0.8	0.9
Rice milk	270	0.4	1.2
Almond milk	372	0.5	0.7
Traditional cow's milk	628	9.0	3.2

Nutritional impact

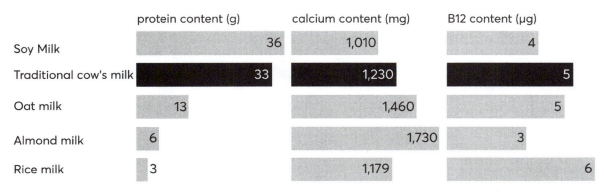

	protein content (g)	calcium content (mg)	B12 content (µg)
Soy Milk	36	1,010	4
Traditional cow's milk	33	1,230	5
Oat milk	13	1,460	5
Almond milk	6	1,730	3
Rice milk	3	1,179	6

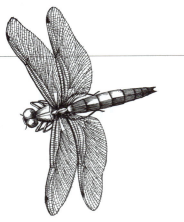

More than 2000 insect species are edible.

Edible Insects

According to the Food and Agriculture Organization, entomophagy—eating insects—may be one solution to meeting global food demand as land mass and biodiversity are lost. Insects are a food staple for 113 countries around the world, with Africa, Asia, and Latin America eating the most. More than 2000 insect species worldwide are edible.

Insects have diverse uses across food systems because they can be eaten directly, added to products to boost protein levels, or included in feedstock mixtures. They can be raised in almost any climate or environment and use less land, feed, and energy than traditional protein sources. They also emit far less carbon. Grasshoppers, for example, produce only trace amounts of greenhouse gas emissions per kilogram compared to livestock.

⊕ **104**

Currently, only 12 crops and five animal species make up 75 percent of global food consumption.

The bugs we're eating

31%
Beetles

18%
Caterpillars

14%
Bees, wasps & ants

13%
Grasshoppers, locusts & crickets

13%
Other groups

10%
Cicadas, leafhoppers & planthoppers

Cricket burger recipe

adapted from Skye Blackburn

Ingredients

- 400g tin chickpeas, drained
- 340g tin sweet corn, drained
- 20g dried cricket or locust powder (can be found online)
- ½ bunch fresh coriander
- ½ teaspoon paprika
- ½ teaspoon ground coriander
- ½ teaspoon cumin
- zest of 1 lemon
- 3 tablespoons flour
- salt to taste

Method

Pick the coriander leaves, then add half into your food processor with all of the stalks. Add the cricket powder, spices, flour, lemon zest, and a pinch of salt. Add the drained chick peas and corn.

Blend the ingredients until roughly mixed. Divide your bug burger mixture into 4 even patties and lightly coat the outside of each with plain flour to avoid sticking.

Refrigerate or freeze to ensure that they hold together when cooked.

Saute in a hot frying pan with olive oil. Cook on one side until brown, then flip and finish cooking.

Excellent with vegan cheese, heirloom pickles, and dijon mustard on a sourdough roll. (It's up to you whether you want to tell your guests about the crickets before or after lunch.)

1 kg of food: Comparing crickets to cattle

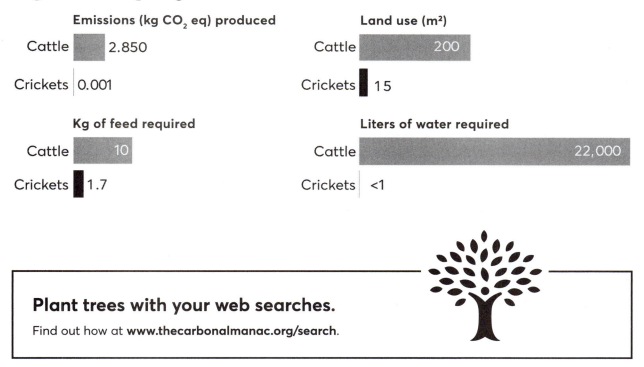

Emissions (kg CO$_2$ eq) produced

Cattle 2.850

Crickets 0.001

Land use (m²)

Cattle 200

Crickets 15

Kg of feed required

Cattle 10

Crickets 1.7

Liters of water required

Cattle 22,000

Crickets <1

Plant trees with your web searches.

Find out how at **www.thecarbonalmanac.org/search**.

Backyard Regeneration

Two approaches to agriculture

The front lawn is a status symbol created in England and imported to the United States. A sweeping, weed-free lawn symbolized to the neighbors that one could afford a staff to keep it in check.

This requires water, chemicals, and fuel.

George Washington, Thomas Jefferson, and even Woodrow Wilson used sheep instead to keep the grass at a manageable height.

> As the environmental and financial cost of a status-filled front lawn becomes more clear, some homeowners are switching to a different approach based on regenerative agriculture.

Regenerative practices keep the carbon where it began before we got here—underground—because it remains captured in the soil. The end goal is restoring and improving the land as well as the entire environment. Practices following this philosophy are more likely to increase biodiversity, enrich the soil, and improve watersheds.

The backyard footprint

An estimated 40 million acres of land are covered by lawn in the US. However, any carbon absorbed by those tiny grass leaves can't compete with carbon emissions from the gas-powered equipment running every weekend to maintain them. Fertilizers present an additional environmental burden—for every ton of nitrogen created to manufacture them, four or five tons of carbon are added to the atmosphere.

However, small changes can make a big impact. Here are three simple steps to begin reducing the carbon footprint of a yard using regenerative principles:

- Let leaves lie and grasses grow. The reduced carbon emissions resulting from one less hour of mowing and blowing per week quickly add up.
- Compost. Stir decomposing leaves, plant and food waste together to create a mix of nutrients to cycle back into the soil.
- Shrink the lawn and add more native plants, shrubs, trees, vegetables, and fruits. The added biodiversity will help the backyard become self-balanced and resilient.

⊕ **108**

George Washington, Thomas Jefferson, and Woodrow Wilson used sheep to keep the White House lawn at a manageable height.

The lawn care crew at the Wilson White House.

Composting

Every year, around $1 trillion worth of food is dumped in landfills, making food the number one material found there, surpassing plastic and paper products. This food waste, covered and in a vacuum, rots. The anaerobic fermentation process that follows emits methane.

Food waste is estimated to be responsible for 8-10 percent of global greenhouse gas emissions every year. Instead of dumping food scraps in landfills, composting food waste could reduce the amount of greenhouse gas emissions by 50 percent.

Composting uses an *aerobic fermentation* process to avoid the release of methane. The practice reduces greenhouse gas emissions and puts carbon back into the soil.

In the US, composting has more than tripled since 2000, with more than 2.3 million metric tons composted in 2018. This is a small fraction of the 35 million metric tons of food discarded annually.

⊕ **260**

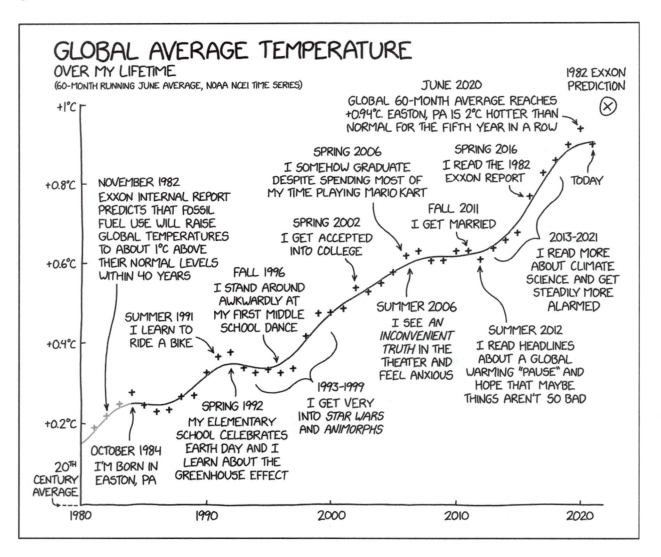

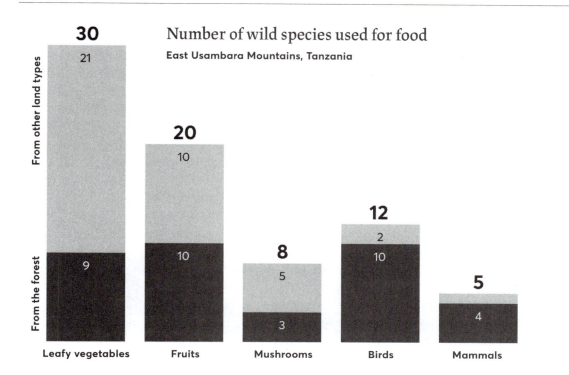

Number of wild species used for food

East Usambara Mountains, Tanzania

From other land types / From the forest

	Leafy vegetables	Fruits	Mushrooms	Birds	Mammals
Total	30	20	8	12	5
From other land types	21	10	5	2	1
From the forest	9	10	3	10	4

Forests Support Food Security

One-fifth (1.2 to 1.6 billion) of the people on Earth depend upon forests for their livelihood, shelter, water, fuel, and food security. Of those, 60 million are Indigenous peoples who depend fully on the forests.

Studies of Indigenous communities in 22 countries in Africa and Asia reveal that they typically consume over 120 different wild foods. Forests have the potential to transform food systems as they provide some of the most nutrient-rich wild foods, including:

· wild meat or bushmeat from birds and mammals
· fresh and saltwater fish
· nuts and seeds
· fruits and berries
· mushrooms
· leafy greens
· insects

Access to healthy wild foods by people living in and near forests adds to food security in stable times but also provides a safety net in the difficult times brought on by climate-related drought and crop failures, as well as war.

⊕ **250**

Countries with the most forested area per capita

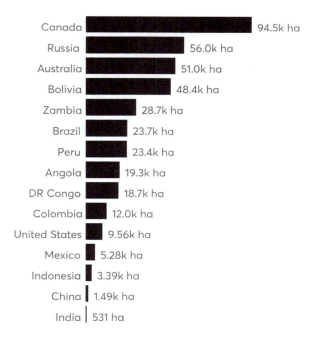

Country	Forested area per capita
Canada	94.5k ha
Russia	56.0k ha
Australia	51.0k ha
Bolivia	48.4k ha
Zambia	28.7k ha
Brazil	23.7k ha
Peru	23.4k ha
Angola	19.3k ha
DR Congo	18.7k ha
Colombia	12.0k ha
United States	9.56k ha
Mexico	5.28k ha
Indonesia	3.39k ha
China	1.49k ha
India	531 ha

Carbon Labeling

Carbon labels are gaining popularity and have begun to appear on food and consumer goods packaging. Some organizations are labeling their entire enterprise. Much like nutrition and Energy Star labels, carbon labeling is a first step in helping consumers make climate-educated buying decisions based on understanding how consumption affects the climate.

The data presented on carbon labels often reflects an estimate of a product's carbon emissions in grams or kilograms of carbon emitted from "cradle to grave." Estimating carbon emissions is a complicated calculation that varies for every product and includes carbon emitted during a product's creation, transportation, use, and end of life.

There are also currently dozens of third-party sustainability accreditations for which organizations can apply and qualify. Each program certifies something different, which requires that the consumer sort out what is implied by the various logos and accreditations.

A first step is recognizing what is implied by the labels:

- **Carbon Neutral:** The organization removes the equivalent amount of carbon it emits. To become carbon neutral, organizations can emit less carbon or offset their carbon emissions by purchasing carbon credits.
- **Climate Positive & Carbon Negative:** The organization removes more carbon from the atmosphere than it emits (it's understandably confusing that they both currently mean the same thing).
- **Climate Neutral:** To achieve this standard, an organization reduces greenhouse gases to zero, while at the same time eliminating all other negative environmental impacts caused by the company.

- **Net-Zero Carbon Emissions:** An activity releases net-zero carbon emissions into the atmosphere.
- **Net-Zero Emissions:** The entire array of greenhouse gases (not just carbon) released is balanced against the total amount of greenhouse gases removed from the atmosphere.
- **Carbon Offset Programs:** Investment in carbon-reducing or storage projects is used to balance an organization's carbon emissions. For example, a company might buy carbon credits that invest in clean wind energy to offset emissions from corporate travel.

There are concerns that some offset programs do not reduce the amount of carbon they claim. However, legitimate offset programs do exist. When deciding on the trustworthiness of a carbon label, how the program offsets its carbon also needs to be understood. In many carbon accreditation programs, companies buy carbon credits to offset their emissions.

Neither the United States Environmental Protection Agency nor the Federal Trade Commission regulates carbon label standards. Until there is widespread and standardized carbon labeling, there is no easy way to compare the climate impact of products in the same category.

⊕ **214**

Small minds are concerned with the extraordinary, great minds with the ordinary.

— Blaise Pascal

Logos adorning consumer products

SUSTAINABLE PACKAGING

Glass used for packaging can be recycled an endless number of times with no decrease in quality. Beginning with used and crushed mixed glass (called cullet), manufacturers can create new glass bottles using substantially less energy than starting from sand, soda ash and limestone. Traditional glass manufacturing requires heating the ingredients to 1700°C/3092°F in a furnace. Melting crushed glass is far easier.

A new technology being tested in France uses electricity instead of natural gas, meaning that the entire recycling process could use renewable energy.

Footprints and Labels

Every decision has consequences.

Using energy releases carbon. Living in the modern world means that each one of us has an impact, and the choices we make can change it dramatically. Our purchases, transportation, and daily habits add up.

> **There can be no Plan B because there is no planet B.**
>
> — UN Secretary-General Ban Ki-moon

The term "carbon footprint" was popularized in 2004, when oil giant British Petroleum launched their "carbon footprint calculator" as a way to encourage people to see climate change as their personal responsibility. While too much effort focusing solely on one family's footprint can distract from the large-scale systemic changes that need to be taken, it's also true that our personal and professional choices can cause widely varying impacts over time.

Some of these include:

- air travel
- method of commuting/choice of vehicle
- diet
- housing

Today, it is possible to estimate the amount of greenhouse gas emissions produced by a given action using a carbon footprint calculator. These calculators enable businesses, individuals, and families to assess the footprint they leave behind when making day-to-day decisions concerning home and office energy use, transportation, and waste production. The Carbon Calculator offered by the United States Environmental Protection Agency is an example of a calculator individuals and families can use.

In addition, companies are beginning to put labels on the products they sell. The challenges here range from the lack of transparency in calculating the impact of the labeled products to the minor impact a specific consumer product has on global carbon emissions. With persistent focus, the labels may become more stringent and more transparent.

There are two ways that calculators and labels could change how the public deals with the climate.

The first is that the act of paying attention to our personal habits opens the door to realizing that we also have a voice in how the systems around us work. It's easier to understand the realities of systemic change once we have personal experience changing our own decisions.

> **The act of paying attention to our personal habits opens the door to realizing that we also have a voice in how the systems around us work.**

The second is that they send a signal that corporations and politicians are very good at listening to data that will impact them in the short run. Once a company sees that a carbon label helps them increase sales, they will work to get more of those labels—or perhaps even better labels. And once elected officials realize that climate is an issue that people are paying attention to, they are far more likely to pay attention as well.

🌐 **212**

Green Steel

The steelmaking industry accounts for 7-9 percent of global CO_2 emissions annually. That's more than the 2019 CO_2 emissions of Japan and India combined.

Why? About 70 percent of worldwide steel production uses coal as fuel in blast furnaces to melt the iron ore. Every metric ton of steel produced emits 1.8 tons of CO_2.

Green steel production doesn't have the same emissions impact. Swedish venture HYBRIT—which delivered its first batch of green steel as a trial run to Volvo in August 2021—uses hydrogen and renewable electricity to melt iron. The hydrogen gas (which they create from water via electrolysis) heats a furnace to 815°C/1500°F, reducing the iron but not melting it. The ore is then liquified with bolts of electricity.

HYBRIT will officially launch its green steel commercially in 2026. It projects that if the entire steel industry went green, annual CO_2 emissions could be cut by 90 percent. This could mean a global CO_2 emissions reduction of 6-8 percent annually.

To be clear, this new technique isn't emissions-free. Like traditional steelmaking, HYBRIT binds the carbon from coal with iron to create its steel. But the CO_2 from that process represents a small fraction of the CO_2 emitted by coal-fueled blast furnaces. This method also consumes plenty of electricity. Melting and shaping the iron needs about 900 kWh of electricity per metric ton of steel. And electrolysis needs 2,600 kWh to produce enough hydrogen gas to produce a metric ton of steel.

> **EXTENDING LIFE EXPECTANCY**
>
> A 2013 study showed that eliminating coal from northern China would extend average life expectancy by 5 years per person. Eliminating cancer from Europe and the United States would extend life expectancy by 3 years per person.

How much CO_2 those power requirements would emit varies from country to country, depending on the electrical grids and power plants feeding the facilities. US plants using the hydrogen method could save 20 percent in CO_2 emissions, while those in the EU are projected to save 40 percent. Sweden's CO_2 savings projections? Nearly 95 percent.

On the other hand, China's electrical grid is so coal-dependent that making green steel there would actually generate 30 percent *more* CO_2 emissions.

Scalability poses a potential obstacle as well. The steel industry would have to build new plants both for hydrogen-based steelmaking and for hydrogen gas production. To reach the 2 billion tons of steel made every year, the HYBRIT method would consume almost 7 trillion kWh of renewable electricity. That translates to 91 percent of *all* the renewable electricity generated in 2020.

⊕ **224**

The steelmaking industry accounts for 7-9 percent of global CO_2 emissions annually. That's more than the 2019 CO_2 emissions of Japan and India combined.

A new law in France requires all ads for cars to include a message that people consider walking, carpooling, riding a bike, or using public transportation instead.

Low-Carbon Concrete

Concrete is the number one source of embodied carbon in buildings. *Embodied* refers to the amount of carbon emitted during construction. Concrete is used in the foundations of homes, cities, bridges, and roadways and accounts for 8 percent of all man-made CO_2 emissions worldwide.

One of the main ingredients in concrete is cement, which is made in limestone kilns by burning coal or natural gas at high heat around 1450°C/2642°F. This combustive process releases approximately one ton of carbon dioxide into the air for every ton of cement produced.

Some of the carbon-emitting steps in the production of concrete buildings include:

- building the plants and machinery for quarrying
- actually quarrying the limestone, sand, and aggregate
- transport from the quarrying site
- heating the limestone
- manufacturing the concrete
- transport to the construction site

Creating lower-carbon concrete

There are two main methods of reducing cement's carbon contribution into the air:

1. Capture and storage
2. Reformulation

In traditional manufacturing methods, airborne CO_2 emissions from cement production can be captured and added back into ready-mix concrete for storage.

Newer methods reformulate cement with similar-behaving materials that generate less CO_2 than traditional manufacturing. For example, cement can be supplemented with carbon-heavy waste (called fly ash) from coal-fired electric generating plants to decrease concrete's carbon footprint. It also increases its strength and workability.

⊕ **213**

CURRENT SOLUTIONS

Blue Planet Systems, a concrete company in Los Gatos, CA, has patented a mineralization process to capture CO_2 from flue gases. They do this by converting the flue gases to carbonate (CO_3)—which locks up carbon and results in lower or negative carbon concrete while promising the same integrity as standard concrete.

US Concrete in Euless, Texas, takes fly ash from burning coal or steel slag and adds it into cement—reducing the amount of cement usage by as much as 50 percent. Additional aggregate from slag lowers the amount dumped into landfills and reduces the need to mine raw materials.

Geopolymer Solutions of Conroe, Texas, produces heat-free concrete made of combined recycled fly ash, granulated slag, and other naturally occurring minerals instead of cement. This lowers carbon emissions by 90 percent compared to using cement in concrete.

Reducing Embodied Carbon in Building Materials

Twenty-three percent of global carbon emissions originate from the production of three building materials: concrete, steel, and aluminum.

Six ways to reduce embodied carbon in construction

1. Repurpose buildings and reuse materials to build them.
2. Invest in the development of low-carbon concrete.
3. Use lower-carbon or carbon-sequestering materials (e.g., sustainably-produced timber).
4. Design structures with fewer finishing materials.
5. Maximize structural efficiency while using minimal carbon-intensive materials.
6. Minimize waste from construction.

Alternative low-carbon building materials

- Bamboo is rapidly renewable and a versatile building material.
- Timber produced from sustainably-farmed forests is a strong structural material, even in multi-story buildings.
- Straw from wheat, rice, rye, and oats can be used for wall infill and thermal insulation.
- Hempcrete is made of the woody inner core of hemp combined with lime and a hardening additive. It can replace clay bricks and cement-based concrete.
- Wool is an excellent material for high-efficiency thermal insulation.

⊕ **229**

> *Each one of us matters, has a role to play, and makes a difference.*
> *Each one of us must take responsibility for our own lives, and above all, show respect and love for living things around us, especially each other.*
>
> — Jane Goodall

Canada's terrestrial landmass holds almost a quarter of the world's carbon stock, most of it in the form of peatland.

Building Materials That Sequester Carbon

Nearly 40 percent of global CO_2 emissions are generated by buildings each year. Of those emissions, just three materials—concrete, steel, and aluminum—are responsible for 23 percent of total global emissions (most of this is used in the built environment).

Changes in technology permit us to create building materials that sequester carbon during their creation rather than emitting it. Carbon sequestration is one process to address the climate challenge. It proceeds by capturing and storing atmospheric carbon dioxide to prevent it from remaining in the Earth's atmosphere.

Buildings can be converted to emission sinks by using biogenic materials that store carbon and reduce emissions during the production process. Biogenic, carbon-storing building materials can be produced from biomass (e.g., annually harvested agricultural residues and purpose-grown fibers), such as rice hulls, wheat straw, bamboo leaf ash, sunflower stalks, hemp, algae, and seaweed. When a building uses plant-based materials in its construction, those materials sequester carbon in the building.

Carbon-storing materials

- **Bioplastic:** Made from biochar—a carbon-rich substance made by burning biomass without oxygen.
- **Mycelium:** An inexpensive biomaterial that forms the root system of fungi, feeds on agricultural waste, and in the process sequesters carbon that was stored in this biomass. Mycelium can be used as a fire retardant and as insulation.
- **Carpet tiles:** Constructed from recycled plastic and various biomaterials, these products can store more embodied carbon than is emitted.
- **Wood:** A fully-grown tree can remove 22 kilograms of CO_2 from the atmosphere each year, making wood carbon negative if responsibly sourced and compensated by new planting.
- **3D-printed wood:** Sawdust and lignin discarded by the timber and paper industries can be converted to a 3D printing filament. Fewer trees need to be cut down and waste wood won't need to decay or be incinerated, which would re-release the carbon that is stored.

- **Olivine sand:** One of the most common minerals on Earth, olivine sand is capable of absorbing its own mass in CO_2 when crushed and scattered on the ground. It is used as fertilizer and a replacement for sand or gravel in landscaping. A carbonated version can be added in the production of cement, paper, or 3D-printing filaments.
- **Concrete:** Certain types of concrete capture carbon in production while substituting emissions-intensive cement with waste slag from the steel industry. Cement is responsible for 8 percent of all greenhouse gas emissions.
- **Bricks:** CO_2 can be injected into industrial waste such as mine tailings, turning it from a gas into a solid that can then be used to create cement bricks and other building materials. The process replicates the same mineral carbonation process that takes place in nature as carbon dioxide dissolves in rainwater and reacts with rocks to form new carbonate minerals.

⊕ **265**

"

Our economic system and our planetary system are now at war. Or, more accurately, our economy is at war with many forms of life on Earth, including human life. What the climate needs to avoid collapse is a contraction in humanity's use of resources; what our economic model demands to avoid collapse is unfettered expansion. Only one of these sets of rules can be changed, and it's not the laws of nature.

So we are left with a stark choice: allow climate disruption to change everything about our world, or change pretty much everything about our economy to avoid that fate. But we need to be very clear: Because of our decades of collective denial, no gradual, incremental options are now available to us.

The bottom line is that we are all inclined to denial when the truth is too costly—whether emotionally, intellectually, or financially. As Upton Sinclair famously observed: "It is difficult to get a man to understand something, when his salary depends upon his not understanding it!"

Renewables are, in fact, much more reliable than power based on extraction, since those energy models require continuous new inputs to avoid a crash, whereas once the initial investment has been made in renewable energy infrastructure, nature provides the raw materials for free.

It is a civilizational wake-up call. A powerful message—spoken in the language of fires, floods, droughts, and extinctions— telling us that we need an entirely new economic model and a new way of sharing this planet.

"

— Naomi A. Klein

Zero-Emission Homes

Before it is even occupied, a house operates at a carbon deficit. In fact, about one-third of the carbon created by housing occurs during its construction.

Constructing, operating, and demolishing buildings uses almost half of the energy produced in the United States, producing many gigatons of greenhouse gas emissions.

> **Building construction and maintenance contribute roughly 39 percent of the world's energy-related carbon emissions.**

The zero-energy standard

The zero-energy standard requires that a house produces as much energy as it uses. This means these homes are airtight, well insulated, and use energy-efficient electric appliances. Heating and cooling do not require carbon-producing oil and natural gas.

As a passive home, renewable energy is amplified by better overall house design. This allows these homes to operate with no additional energy bills or carbon emissions and be comfortable to live in as well. Zero-energy homes are functional in cold and warm climates and are often indistinguishable from traditional homes except for their lower energy usage and costs.

According to the Zero Energy Project, a typical home built to the following standards will cost about 10 percent more than a traditional house. But the reduced energy costs will result in dramatically more savings than the increased mortgage payments, making the house cheaper to own in the long run.

🌐 **111**

The Zero Energy Project provides recommendations for building and remodeling to this standard:

1. Work with an architect or builder experienced in zero-energy homes.
2. Orient the building to maximize winter sun and summer shade.
3. During the design phase, use software to model and optimize the home's future energy usage.
4. Ensure windows and doors are airtight to reduce energy use for heating and cooling.
5. Invest in significant insulation.
6. Use triple-paned windows and highly insulated doors.
7. Create a ventilation system to add fresh filtered air and control moisture.
8. Select an energy-efficient heating and cooling system like a ductless heat pump.
9. Use the latest technologies to minimize water use and heat water efficiently.
10. Install LED lighting and strategically place windows to maximize natural light.
11. Select energy-efficient appliances and electronics.
12. Harness the sun for renewable energy by installing grid-tied solar roof panels.

RETURNED MERCHANDISE

Because it can be cheaper to throw away merchandise than to repackage, re-inventory, store it, resell it, and ship it out again, as much as five billion pounds of returned merchandise end up in US landfills every year.

Cross-Laminated Timber

Instead of steel or concrete, it's now possible to reliably build multi-story buildings with wood.

Cross-laminated timber (CLT) uses dried lumber stacked in alternating directions held together with glue. By pressing this wood into panels and beams, it's possible to build high-strength, reliable, and fire-resistant structures. CLT is commonly made with formaldehyde-free polyurethane or EPI.

Building with wood

Wood is a renewable resource and a resilient building material. Wood structures can be safely built to comply with building codes. In terms of energy usage and air pollution, wood outperforms other building materials such as concrete or steel and can be easily adapted or re-used with basic tools.

⊕ **222**

CROSS-LAMINATED TIMBER (CLT)

CLT panels are fabricated off-site, making the process cheaper and better for the environment.

Quick on-site erection that is faster than working with steel and concrete.

Wooden buildings are lighter, requiring a shallower foundation.

Wood sequesters carbon.

CLT is fifteen times more thermally efficient than concrete, reducing building energy demands.

Carbon cycle in the built environment

At every step of the building process, carbon emissions and extracting resources from nature occur.

11 percent of embodied carbon emissions takes place during the building of a home, including:
· harvesting raw materials
· demolition of existing structures
· transportation of laborers and materials to and from the job site
· manufacturing components like windows, doors, and paint
· assembling the structure

28 percent come from operational emissions that include the energy needed to:
· power the systems and appliances
· heat and cool the home

CO_2

Resource extraction from nature

Manufacturing

Nongreen construction

Building usage

Recycling

Disposal

Green construction

Solar-powered green home

CO_2

Wool & Hemp: Allies in Construction

Hemp grows fast, requires no herbicides or pesticides, and adapts to most climates. It absorbs more carbon per hectare than trees, and in 120 days produces the same usable biomass that softwood produces in 120 years. Hemp cultivation restores soils and offers a productive crop rotation alternative for farmers who lack suitable machinery. Hemp is also great for construction. When mixed with lime, it becomes "hempcrete" and can be used to build walls as well as replace concrete in non-load-bearing structures. Hemp boards can be used instead of plywood or other types of boards that may contain chemicals.

Additional benefits of hemp products when compared to traditional building materials include:

- carbon negative (absorbs carbon instead of emitting it)
- efficient insulating material
- fire-resistant
- recyclable
- lightweight

GLACIAL MELT

Every year for the past 20 years, the planet's glaciers lose about 267 gigatons of mass. That volume of water is enough to cover the entire country of Ireland in 3 meters of water annually. The loss is currently accelerating at a rate of 48 gigatons per year increase per decade.

Sheep eat plants. The plants capture carbon, which the sheep use to grow. Woolmark reports that 50 percent of the weight of sheared wool is carbon.

Like hemp, wool is ideal for construction. It can be used as thermal insulation for buildings. It insulates even as it absorbs and releases moisture. Wool also improves air quality by trapping chemicals like formaldehyde (a harmful gas found in building materials), nitrogen oxide, and sulfur dioxide.

Both wool and hemp sequester carbon and offer the following advantages for construction:

- Natural and non-toxic
- Biodegradable and compostable
- Naturally fire-retardant and antimicrobial
- Strong and long-lasting
- Sound absorbent

⊕ 235

We are our choices.

— Jean-Paul Sartre

Green Building Certifications

LEED and similar programs or assessments verify how sustainable building projects are. These programs have created a competitive dynamic that prompts architects, builders, and developers to push for the highest-possible ratings. Green building features that factor into these assessments include:

- Energy and water efficiency
- Use of renewable energy
- Reduction of waste and pollution
- Consideration of indoor air quality
- Use of sustainable, non-toxic materials
- Positive environmental design, construction, and operation
- Designs that adapt to their environment

However, critics say the checklist approaches of LEED and BREEAM can result in buildings that are actually not energy efficient.

LEED: Leadership in Energy & Environmental Design

- Created in 1998 by the nonprofit US Green Building Council as a rating system for sustainability.
- The highest LEED certification rewards building projects for features such as reduced parking, being located near safe biking options, and measuring potable water consumption. The cost of being evaluated ranges from $5,200 to more than $1 million.

BREEAM: Building Research Establishment Environmental Assessment Methodology

- Set the world's first sustainability standards in 1990 for the evaluation of UK building projects in any phase (from development to refurbishment).
- More than 591,000 buildings in 90 countries have been BREEAM-certified after being assessed on their design, construction, and proposed use.

DGNB: Deutsche Gesellschaft Für Nachhaltiges Bauen (German Sustainable Building Council)

- Established in 2009, the Germany-based DGNB differs from LEED and BREEAM because it also evaluates building projects on their social, political, and economic relevance.
- As of January 2020, the DGNB had certified 5,000 projects in 29 countries.

🌐 247

MICROPLASTIC MENACE

Microplastics are small fragments of plastic less than five millimeters in size that are a significant threat to the environment and human health. In recent studies, tiny bits of plastic in the atmosphere have been shown to absorb infrared light, contributing to climate change.

I was hoping it wouldn't take this to convince you of climate change.

BIZARRO.COM

What Is Carbon Offsetting?

Climate change isn't local. It doesn't matter if the CO_2 emitted is in Manila—it will have an impact everywhere in the world.

Carbon offsetting is based on the idea that someone can cancel out the impact of their carbon emissions by paying someone else to take a similar amount of carbon dioxide and other greenhouse gases out of the atmosphere.

How it works

Carbon offset businesses sell credits that represent units of CO_2 reduction or removal so people or groups can counteract any emissions their actions have caused.

Offset types include:

- **Forestation:** Because trees in some locations sequester CO_2 so efficiently, restoring depleted forests, creating new forests, and conserving existing ones are popular CO_2-reducing options. *The Carbon Almanac* project itself will replace every tree used for its print runs by planting 10 new ones.
- **Renewable energy funding:** By making wind power, solar power, hydroelectricity, nuclear energy, and biofuel cheaper, fewer fossil fuels will be burned.
- **Carbon or methane capture:** These technologies remove GHGs from the atmosphere and store or transform them.
- **Energy conservation funding:** These projects offsets new emissions by decreasing the general demand for energy (e.g., energy-efficient buildings using LED bulbs and green materials).

Compliance market vs. voluntary market

Two separate markets exist—one for entities legally bound to keep their CO_2 emissions under specific numbers and the other for people and companies voluntarily reducing their carbon footprints.

Compliance market: Overseen by various regulatory groups such as the UN's Clean Development Mechanism (CDM), this market caters to nations and companies mandated to cap their annual emissions by pacts like the 2015 Paris Agreement.

For instance, the CDM operates a "cap-and-trade" system meant to make sustainability efforts more accountable for participating countries. Each nation gets a specific CO_2 cap and a number of emissions permits. To stay under the cap and avoid penalties, they can:

- cut emissions
- buy offsets from within the compliance market to meet their cap
- trade with countries that have excess emissions permits

Voluntary market: Estimated to be worth $50 billion by 2030 this public market is mostly unregulated. But over time, globally recognized standards have been set by independent certification groups, some of which keep public records of active and retired offset projects.

Both types of markets seek to use market forces to produce efficiencies, transparency, and a simple metric—the same forces that produced much of the problem to begin with, but in reverse.

Carbon offsetting done well

To ensure the integrity of this practice, an offset must:

- **Be real and measurable:** One credit should correspond to one ton of atmospheric carbon dioxide (or its equivalent in other GHGs) that is reduced, avoided, or otherwise removed.
- **Be long-lasting:** It's tempting for an entity to sell a carbon offset today and then chop down a forest tomorrow. The general convention is that captured CO_2 should remain in place for about 100 years.
- **Offer incremental gains:** If an offset offers a CO_2-reducing action that would have happened anyway, it shouldn't be certified.
- **Be unique:** A credit can't be applied more than once. Once retired, it can't be resold. This is a difficult measurement and enforcement standard—one that organizations are working hard to quantify and certify.

Caveats

Critics point out that offsetting allows users of fossil fuels to continue to avoid confronting the climate emergency of carbon combustion. Also, some companies have "greenwashed" their destructive environmental impacts with splashy offset investments.

And as with any market, carbon offset fraud exists, potentially to the tune of millions of useless credits. Currently, there's no universal regulator to verify offsets. Red flags companies should watch out for include:

- unrealistic projections and very low prices
- tree-planting efforts in areas that don't suffer from deforestation
- offsets that don't articulate how they offer *additional* CO_2 reduction or removal
- projects causing dislocations and human rights violations

⊕ **348**

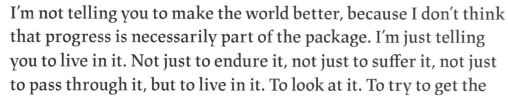

I'm not telling you to make the world better, because I don't think that progress is necessarily part of the package. I'm just telling you to live in it. Not just to endure it, not just to suffer it, not just to pass through it, but to live in it. To look at it. To try to get the picture. To live recklessly. To take chances. To make your own work and take pride in it. To seize the moment.

And if you ask me why you should bother to do that, I could tell you that the grave's a fine and private place, but none I think do there embrace. Nor do they sing there, or write, or argue, or see the tidal bore on the Amazon, or touch their children. And that's what there is to do and get it while you can and good luck at it.

— Joan Didion

EMISSIONS FROM FLYING

Each passenger taking a roundtrip flight between the Caribbean and Germany produces four metric tons of emissions—the same as 80 residents of Tanzania do in an entire year.

Clay Dumas of Lowercarbon Capital points out that carbon capture is undergoing a technology boom, with new companies and funding frequently being announced. Some of the companies to investigate are: Charm, Verdox, Running Tide, Eion, Mission Zero, and Sustaera.

Direct Air Capture

Direct air capture (DAC) is a process for removing carbon dioxide from Earth's atmosphere in an effort to reduce its effect on climate change. DAC involves using powerful turbine fans to suck carbon dioxide from the air and then store it or reuse it.

CO_2 is the biggest contributor to the greenhouse effect and global temperatures rising. DAC seeks to reverse the process of industrialized emissions. A DAC system forces air in the atmosphere to pass over a liquid solvent or a solid absorbent filter (called a sorbent), which absorbs CO_2. It then releases the remaining air.

The liquid solvent or solid sorbent is then heated to release the CO_2. Liquid solvent-based systems are energy-intensive and require very high temperatures of nearly 900°C/1562°F to release CO_2. Solid sorbent-based systems need to be heated to only 80°/176°F for the CO_2 to be released. It's then captured and stored, and the solvent or the sorbent is now ready for reuse.

The CO_2 gas that has been captured is then injected underground for storage in certain geological formations. After CO_2 has been pulled from the air, it can be mixed with water and pumped underground, where it reacts with bedrock to form carbonate minerals. Storing carbon dioxide this way removes it completely from the carbon cycle. This is called negative emissions.

Alternatively, the captured carbon dioxide can be used in industrial applications, such as hardening concrete or producing synthetic fuels. This means the CO_2 could be trapped inside concrete for many years, or burned and returned to the atmosphere.

Synthetic fuels could be deemed carbon-neutral because they simply return the carbon to the atmosphere shortly after it is captured. But there is an energy cost to this cycle.

Organizations working on this problem are trying to improve scale, lower cost, and create a resilient methodology. Heirloom lists these objectives for high-quality DAC:

- **Durable:** Captured CO_2 should be stored away for as long as possible, preferably for thousands of years.
- **Additional:** There should be additional CO_2 removed above and beyond what would have happened in the business-as-usual scenario.
- **Timely:** Removing CO_2 today is preferred to removing it tomorrow in order to avoid climate tipping points like ecosystem collapse or ice sheet loss.
- **Sustainable:** Land, water, feedstock, and energy usage should be minimized to ensure the process is truly regenerative and non-extractive.
- **Net-negative:** Cradle-to-grave emissions should be well understood so we can accurately calculate how much net CO_2 is actually being removed.
- **Monitorable:** The energy and emissions at each step of the process should be continuously monitored to ensure that CO_2 is efficiently captured and contained within the system.
- **Renewable:** All systems should operate on as much renewable energy as possible.
- **Resilient:** Equipment and operations must be designed to be resilient and adaptable to changing weather and climate.
- **Safe:** Solutions must pose little to no risk to the health of workers, local communities, and surrounding ecosystems.

A magic bullet?

Analyses of net zero by 2050 goals propose that DAC will grow dramatically, projecting increased capacity by a factor of 20,000 in just 10 years—from a few small plants to 85 megatons a year. Industries like air transport, which aren't able to cut their emissions, are asserting that the CO_2 they create will be relieved by the increased prevalence of DAC.

There are only 19 DAC plants in operation today. The largest plant under development in the US will capture one megaton of CO_2 each year and is expected to be operational by 2024.

For context, it takes the world's cars *less than three hours* to produce more carbon than this large DAC can remove in a year.

The CO_2 unit could also require up to 24.7 km² of land to support it. To physically scale the current technology to be able to absorb current levels of emission simply isn't possible.

Challenges

- DAC requires power. The generation of power is causing much of the carbon dioxide problem.
- DAC is difficult to scale.
- DAC is an active, not passive, process. When energy and effort stop being expended, carbon capture ceases.
- Removing carbon from the air isn't as resilient or productive as not releasing it in the first place.

🌐 **253**

Storing Carbon Naturally

Actively working to store carbon is called sequestration. Before humans arrived on Earth, there were two ways carbon was naturally stored: biological and geological.

Biological sequestration occurs when plants absorb the carbon dioxide in the air and convert some of the CO_2 into oxygen and glucose. This process is called photosynthesis. The plants in the ocean do the same, and the water itself dissolves a portion of the carbon dioxide. The Earth stores the carbon in trees, soil, and the ocean. When humans work to create more forests, they're creating the conditions for this sort of storage.

After carbon is stored in this way, it sometimes migrates to the slow cycle for long-term storage.

Geological sequestration is the process of carbon and carbon dioxide being stored as fossil fuels such as oil, gas, and coal. Originally, this took millions of years, which is why they're called "fossil fuels."

When humans release more carbon than biological and geological processes can absorb, the amount of carbon dioxide in the air increases. This shift is a main cause of climate change.

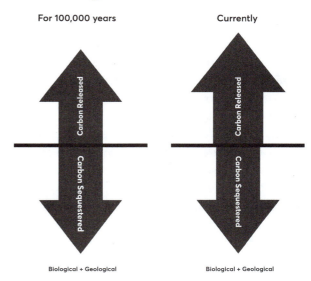

Coastal wetlands, known as "blue carbon" due to their proximity to both sea and land, are being targeted for rehabilitation due to their carbon sequestration potential. Salt marshes, mangrove forests, and seagrass meadows store more carbon than they release.

🌐 **107**

Replenishing Forests

Wood is made of carbon, and a cube of wood about one meter on each side holds a metric ton of CO_2. Because of this, forests play a critical role in counteracting CO_2 emissions. Every year, trees absorb about 2.6 gigatons of CO_2 worldwide. That's about 7.6 percent of the CO_2 emitted by global fossil fuel combustion in 2019.

But the destruction of forests is on the rise. In 2020, forests were destroyed at a seven percent higher rate than in 2019. Rainforests in particular were destroyed at an even higher rate of 12 percent; the Brazilian Amazon alone lost tree cover at a 15 percent higher rate.

Two widespread agricultural practices have driven much of this damage:

- **Deforestation:** Forests are cleared entirely so the land beneath can be put to a different use, such as farming or raising cattle.
- **Degradation:** Illegal or improper logging strips forests of their best trees, ruining vegetation, undergrowth, and soil.

Each has a twofold impact on climate change:

- **Immediate release of CO_2:** The trees that are removed or burned release previously-stored CO_2 back into the atmosphere. The soil beneath the trees emits carbon as well. In 2020, soil in the US stored approximately 50 percent of the forest carbon.
- **Elimination of future carbon storage:** Those same trees and soil are no longer available to remove CO_2 from the air—now or in the future.

The UN has set a goal of increasing forests worldwide by three percent by 2030. This remediation comes in three basic methods:

- **Afforestation:** Establish forests where they never existed before or where they haven't existed for at least 50 years. Fast-growing tree species may yield the best CO_2-absorbing results, and using a wide variety of species can help preserve a region's biodiversity.
- **Reforestation:** Tree planting restores recently cleared forests. Diversifying species is again

1 ton of CO_2

1 ton of CO_2 equals a cube of wood about one meter on each side about the size of a female emperor penguin.

critical, as is choosing those that will thrive under the current conditions of a region versus defaulting to what once grew there.

- **Natural regeneration:** This technique is specific to degraded forests. Regrowth is nurtured on recently cut stumps. These sprouts can then access the support of the removed trees' larger root systems. The land can also be reseeded by the remaining living trees.

Up to two billion hectares of degraded forest land—equivalent to about two-thirds the land mass of Africa—could be viable for natural regeneration. Compared to the other two methods, regeneration costs less and eliminates the need for CO_2-emitting steps like tree transportation. But achieving the densities required for impactful CO_2 absorption has been tricky.

It's unclear how effective any of these methods would be. Claims that planting one trillion trees would remove 25 percent of the CO_2 in the air have been challenged by researchers. And replenishing the planet's tree density is a slow, time-consuming process. In 2020, the UN reported the world was not on pace to meet its three percent goal.

🌐 220

The Limits of Reforestation

Reforestation is the process of replanting trees in areas that were previously forests. Large reforestation projects include Trillion Trees, China's Great Green Wall, Eden Reforestation Projects, and the African Forest Landscape Restoration Initiative.

Reforestation projects receive broad support from governments, businesses, and individuals. Part of the broad appeal is the low monetary cost to support such projects. An individual can pay $1 to have a tree planted. For $3-5 per metric ton, a business can buy a carbon credit to offset emissions.

So strong is the support for trees that afforestation projects are also being presented as a climate solution. This is where trees are grown in areas that historically have not had trees, such as the Sahara Desert.

Simply planting more trees is not necessarily a win-win solution.

Trees are complex organisms, and not all forests are created equal when it comes to carbon absorption. Equatorial forests, particularly coastal mangrove forests, are significantly more effective in absorbing carbon than upland forests in temperate climates.

Projects that focus on quickly planting lots of one type of tree—a monoculture—actually reduce the potential carbon that could be sequestered compared to allowing the forests to naturally regrow. Fast-growing invasive species can overtake native plants and produce more carbon than they absorb. Such forests also reduce biodiversity.

How long the trees last is an important consideration for reforestation. China's Great Green Wall has been discussed over the last 25 years. Questions about quantity versus quality, local wildlife, and tree durability persist. The 2021 forest fires in the US wiped out carbon offsets purchased by companies such as Microsoft and BP.

Critics of reforestation argue that it disguises the need to reduce actual emissions. The amount of land required is also unfeasible.

In order to plant enough forests to absorb the carbon produced by humanity in 2050, an area five times the size of India would need to become a forest.

Reforestation actions may further marginalize vulnerable communities and displace Indigenous peoples if land is taken away and designated for tree planting.

Reforestation can help, but conservation is better

While reforestation done properly can have positive benefits, the main problem is if it detracts from conservation of existing forests. Old-growth forests with many species of trees can store more carbon than new growth. One hectare of old-growth can sequester 100 tons of carbon annually; the same amount of land devoted to new growth only sequesters three tons of carbon annually.

Peatlands, mangroves, old-growth forests, the Amazonian canopy, and marshes also are seen as storing irrecoverable carbon, in that they store so much carbon, their destruction would emit carbon in amounts far more than could ever be offset or made up for. 🌐 **219**

Blue Carbon

Algae, seagrasses, mangroves, salt marshes, and other plants in coastal wetlands absorb and trap carbon as they grow. "Blue carbon" refers to the way coastal and marine ecosystems capture and retain carbon dioxide. Half (or more) of the carbon trapped on the sea floor comes from these coastal forests. They can capture carbon dioxide four times faster than a traditional forest because much of the carbon goes several meters deep into the wet soil. Trapping carbon this way removes it from the atmosphere and reduces the overall level of carbon dioxide in the air.

One hectare of mangrove forest can trap as much as eight tons of carbon dioxide every year, which is more than what a hectare of tropical forest can capture.

🌐 **251**

Over the last half-century, between 30 and 50 percent of the world's mangroves have been destroyed.

Carbon stores by biome

Megatons CO$_2$ per hectare

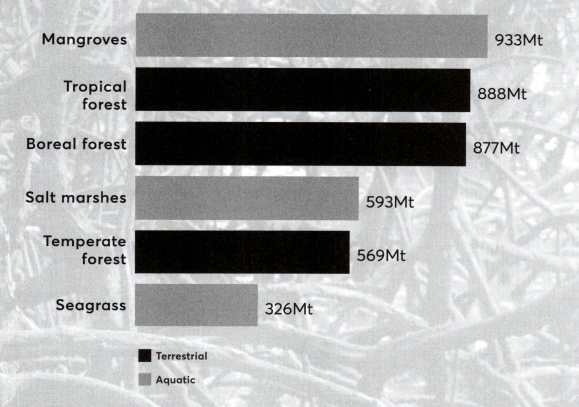

Mangroves	933Mt
Tropical forest	888Mt
Boreal forest	877Mt
Salt marshes	593Mt
Temperate forest	569Mt
Seagrass	326Mt

■ Terrestrial
■ Aquatic

Using Soil to Store Carbon

Soil is alive. Dirt becomes soil when it's inhabited by countless microorganisms that turn it into a vital substrate for the growth of plants.

Soil also stores a lot of the world's carbon in a substance called *soil organic matter* (SOM). The term *organic* here doesn't refer to the absence of chemical fertilizers or pesticides—it refers to the significant amount of carbon present. Typically SOM is 50 to 60 percent carbon. Most soils used for agriculture contain three to six percent SOM.

When plant materials (such as leaves or stems) die and fall to the ground, they're broken down by microbes in the soil. This process turns the plant matter into carbon and creates SOM. The carbon becomes fixed in the soil, preventing it from being released into the atmosphere as carbon dioxide.

Plowing destroys SOM and the storage of carbon in the soil. When farmers till the soil, SOM is brought to the surface. It becomes more available to microbes, which consume the SOM quickly, releasing carbon dioxide into the atmosphere.

Every year, about one to two gigatons of carbon that was stored in soil is released to the atmosphere as carbon dioxide, either due to tilling, erosion, or climate-related changes to soil, such as thawing permafrost.

SOM can be retained or rebuilt, allowing atmospheric carbon dioxide to return to the soil long-term. SOM increases when farmers apply manure, when they leave plant waste (such as corn stalks) in the field to decompose, or when they grow cover crops. Cover crops are planted after the growing season when fields would otherwise be left bare. They're often grasses or clovers that have deep, soil-penetrating roots. When the cover crops are allowed to decompose in the field before a new commercial crop is planted, they meaningfully increase the SOM and carbon in the soil.

Minimizing tilling (called *conservation tilling*) is another way to prevent loss of SOM (or to allow it to regenerate over time). One approach, called no-till planting, uses a specialized planter to place seeds into a small area of loosened soil, eliminating the need to till the entire field.

⊕ **254**

Restoring Soil Health

Dirt is not all the same. Over time, the content of soil changes based on how it has been treated and the environment it is exposed to.

One-third of the world's soils have degraded to the point where they can offer little support to plant or animal life. Some of the causes are:

- tilling
- over-grazing by cattle
- slash-and-burn tree and plant removal
- failure to plant cover crops in winter
- insufficient mulching

Large industrial-sized farms in Asia, Europe, and North and South America contribute to soil erosion by increasingly growing commodities like soybeans, wheat,

rice, and corn. The economic pressures of markets and debt make sustainable agricultural practices challenging to implement in the short run.

Soil health has far-reaching impacts, from the quality of food produced to the amount of carbon in the atmosphere. When soil is healthy, it balances the water cycle and acts as a shock absorber to prevent flooding and erosion. The Dust Bowl in the western United States in the 1930s and flooding in Puerto Rico in 2017 are examples of the catastrophic impacts of climate shifts and natural disasters related to soil health. These changes can have significant impacts on agriculture.

According to the US Department of Agriculture, there are four ways farmers can create better soil:

Minimize Disturbance

- limit tillage
- optimize chemical input
- rotate livestock

Maximize Soil Cover

- plant cover crops
- use organic mulch
- leave plant residue

Maximize Biodiversity

- plant diverse cover crops
- use diverse crop rotations
- integrate livestock

Maximize Presence of Living Roots

- reduce fallow
- plant cover crops
- use diverse crop rotations

On a local level, citizens can vote for legislation and policies in favor of sustainable farming practices, as well as purchase products from sustainable farming operations.

Homeowners can also improve soil health around their properties by diversifying plant species grown year-round and letting natural processes take hold. This maximizes active root systems and creates more biodiversity.

⊕ **105**

How healthy soil can balance the water cycle

ORGANIC Organic matter	**Healthy soil** Animals and plants can breathe and move about easily.	**Unhealthy soil** Water is not absorbed, and it runs across the surface, eroding the soil.
SURFACE Organics mixed with mineral matter	Water is absorbed and stored in air pockets in the soil.	
SUBSOIL Mixture of silt, sand, or clay	When these pockets are filled and the soil becomes saturated, extra water flows through to return to bedrock aquifers.	Animals and plants beneath the surface have limited movements and root growth, and are devoid of water to survive.
SUBSTRATUM Parent rock		
BEDROCK Unweathered parent material		

Geoengineering

If you build a campfire or poorly dispose of an air conditioner, you're changing the environment with your actions. But when companies and countries change the environment intentionally at a large scale, it's called geoengineering.

Geoengineering tactics sound like they're part of the plot of a science fiction movie: deploying solar shields into space to redirect sunlight or sucking CO_2 out of the atmosphere and sending it underground to turn into stone. Scientists are exploring more of these large-scale ways of tinkering with Earth's systems to cool the planet, but so far many are cost-prohibitive, controversial, and full of risks.

Consider solar shields. Though they sound like solid sheets of metal, they'd actually mimic what a massive volcanic eruption does when it spews clouds of ash and chemicals into the air, thereby blocking the sun. It could be possible to put a chemical into jet fuel so that high-flying jets could spread it into the upper atmosphere.

Supercomputers project that reflective sulfur particles sprayed into the stratosphere this way could have a cooling effect. Of course, they'd also impact rainfall, snowfall, and seasonal temperatures. To what extent is unclear, and if the weather changes too drastically, there's no easy way to undo the damage, and everyone suffers. Even if the spraying could be reversed, halting such a program could be dangerous because of a sudden rise in global temperatures and greenhouse gases due to the now-unblocked solar rays.

As for siphoning CO_2 directly from the air and storing it in rock formations underground, 19 plants in Europe and North America already do this at a rate of about 0.01 megatons of CO_2 per year. No one knows how long CO_2 can safely be sequestered this way. If CO_2 escapes containment, soil, water, and air could be tainted, and collecting gas underground could set off tremors and earthquakes. Regardless, to succeed, the process must become cheaper and more efficient—it currently costs up to $600 per ton—because dramatically more carbon-capturing plants would be needed to get closer to removing the thousands of megatons of CO_2 annually produced to achieve net zero by 2050.

Instead of storing CO_2 underground, iron fertilization is an ocean-focused option. By injecting iron sulfate into the water, this process triggers algal blooms which could absorb CO_2 and then sink to the seafloor. Success rates have been scattershot, with anywhere from 5 percent to 50 percent of the blooms descending far enough to have an impact. Complete effectiveness could come at a price however: excess algae might trigger spikes of toxic phytoplankton growth, and storing CO_2 in the ocean could speed up its acidification.

Geoengineering is a risky bet. Some scientists say its impact on global temperatures would be minimal, especially given the high chance of unwanted consequences by doing nothing. Others have pointed out that relying on a quick industrial solution could distract people and corporations from the real work of reducing their CO_2 emissions or ending fossil fuel use.

There are thousands of companies and countries that can unilaterally engage in geoengineering. Expect these experiments to unfold on separate tracks around the world.

⊕ **240**

Geoengineering with Sulfur Dioxide

Some engineers are proposing a cheap and fast approach to slowing climate change—to "take the edge off" while we get our carbon house in order.

Just as a mirror reflects light and a black driveway gets hot on a summer day, the amount of light the outer atmosphere reflects from the sun can have an impact on the temperature of the entire planet.

Thirty years ago, Mt. Pinatubo in the Philippines erupted, creating the worst volcanic eruption in 100 years. The resultant ash created a startling impact: The average temperature on Earth dropped about .5°C/1°F for an entire year. By causing Earth's atmosphere to reflect sunlight instead of absorbing it, the planet became cooler.

Geoengineers are focused on taking this idea and creating an intentional version of a solar shade around the Earth. By choosing different chemicals and spreading them in the high atmosphere via specially outfitted jumbo jets, they hope to change Earth's reflectivity for years at a time, artificially lowering the average temperature on the surface.

Through geoengineering, the natural effect of volcanic eruptions is replicated by adding microscopic particles to the atmosphere. These stratospheric aerosol injections:
- scatter sunlight
- make the sky a bit whiter
- reflect part of the sun's heat
- make the earth a bit cooler

The planetary albedo (reflectivity) can be increased by injecting the atmosphere with sulfur dioxide (SO_2), titanium, or other chemical or mineral substances.

Solar geoengineering treats the symptoms of climate change by changing the earth's radiation balance. The science that studies this is called stratospheric aerosol modification (SAM).

It's estimated that the annual cost of this approach is less than $10 billion—a tiny fraction of most interventions involving climate change. Some experts argue that it could be done with a few hundred planes and begin sooner than many expect.

In 2006, researcher Mark Lawrence noted that "serious scientific research into geoengineering possibilities, such as discussed in the publications by *Crutzen* and *Cicerone*, is not at all condoned by the overall climate and atmospheric chemistry research communities." But by 2016, he concluded, "In the 10 years since these publications, although climate engineering remains a very controversial issue, the sense of a taboo has largely disappeared in the broader Earth sciences research community."

There are a number of untested and real questions about an approach like this:
- How will the chemicals interact with the ozone layer?
- Which countries will regulate this process and how will decisions be made about the location and amount of intervention?
- What will keep organizations and nations from doing this unilaterally? What if one country wants things to be warmer, or one billionaire wants to be famous?
- What will be the effects on the health of humans, animals, plants, and the oceans?
- Are we prepared to do it forever? If not, how will we find the resolve to stop once we're hooked on a relatively cheap and fast solution?

⊕ **259**

THE PLANETARY ALBEDO

Just as a mirror reflects light and a black driveway gets hot on a summer day, the amount of light the outer atmosphere reflects from the sun can affect the temperature of the entire planet.

Whose Job Is It?

The roles of government, business, and individuals in creating change

Glasgow Breakthrough Agenda

During COP26 (the UN Climate Change Conference 2021) 42 world leaders, whose nations represent 70 percent of global GDP, announced the Breakthrough Agenda to reduce greenhouse gas emissions. They pledged to work together to meet the goals of the Agenda.

The Breakthrough Agenda is a global clean technology plan focusing on the five sectors of the global economy responsible for over 50 percent of global greenhouse gas emissions. The plan includes building a coalition of leading public and private initiatives and sharing information about what will help to promote success. The plan is to significantly reduce emissions by 2030.

The five Breakthrough goals are:
Power
Clean power is the most affordable and reliable option for all countries to meet their power needs efficiently by 2030.

Road transport
Zero–emission vehicles are the new normal and accessible, affordable, and sustainable in all regions by 2030.

Steel
Near-zero emission steel is the preferred choice in global markets, with efficient use and near-zero emission steel production established and growing in every region by 2030.

Hydrogen
Affordable renewable and low carbon hydrogen is globally available by 2030.

Agriculture
Climate-resilient, sustainable agriculture is the most attractive and widely adopted option for farmers everywhere by 2030.

Strengthening international cooperation towards solving carbon emissions problems in these five high-impact areas and keeping them at the top of the international political agenda is the aim of the Breakthrough Agenda.

To meet the goal of reducing carbon emissions in these five areas, the signatories agree to contribute to international collaboration that:

- Supports alignment in policies and standards
- Motivates R&D efforts in environmentally-friendly technology
- Increases coordination in public investment among the international community
- Mobilizes private finance to promote these efforts.

Each nation can sign on to some or all of the Breakthrough goals. Some nations signed on for all of the goals; others signed on for only one or two.

The UK will lead an annual Global Checkpoint Process to track and review progress towards rapid transition. The five Glasgow Breakthrough goals include a set of metrics that countries will use as measures to report progress on reaching their commitments.

🌐 128

What is the UNFCCC / Kyoto / Paris Agreement?

What is the UNFCCC?
The United Nations Framework Convention on Climate Change (UNFCCC) was created in an attempt to come together as a global community to address the challenge of global warming.

All nations are impacted by climate change. Some, like island nations, or those with fewer resources to adapt via infrastructure, may feel these impacts more than others.

And not all nations are equally responsible for contributing to the problem in the first place. But climate change is a global problem that requires global cooperation.

At its conception in 1992, the goal was to find equitable ways to reduce emissions in more developed countries, while providing support to developing countries to grow sustainably. Under the UNFCCC, several principles were agreed upon:

- Action should be taken to prevent harm, even where scientific uncertainty remains.
- Parties should act "on the basis of equity and in accordance with their common but differentiated responsibilities and respective capabilities."
- Developed countries should take the lead.

The UNFCCC has near-universal membership—197 countries have ratified the Convention. These countries send delegations annually to multiple meetings culminating each year in a Conference of the Parties (COP). Decisions are reached by consensus, and countries most often negotiate in groups. The Glasgow COP, held in 2021, was the 26th such meeting.

The Kyoto Protocol and the Paris Agreement

The UNFCCC has two main subsidiary agreements: The Kyoto Protocol (KP) and The Paris Agreement.

The Kyoto Protocol, signed in 1997, aimed to control emissions of greenhouse gases (GHG) in a way that reflected differences in nations' economic development and capacities.

For its first commitment period (2008-2012), a list of 36 "Annex I" countries—countries with developed or growing market economies—committed to a GHG emission cap. All 36 countries complied with the Protocol, although nine of them had to compensate for their emissions by funding emissions reductions in other countries.

A second commitment period from 2013-2020 was agreed to in 2012 but has not entered into force.

At the same time that the second commitment period of the KP was being negotiated, another conversation was underway, which eventually led to the Paris Agreement.

The Paris Agreement was adopted in 2015. Its main goal is to limit the rise of the mean global temperature to 2°C/3.6°F above pre-industrial levels (preferably limiting it to 1.5°C/2.7°F). The Paris Agreement differs significantly from the Kyoto Protocol in that it requires **all parties** to develop "nationally determined contributions" (NDCs) and to report regularly on emissions and progress on implementation.

In an NDC, each country describes the actions it will take to reduce GHG emissions, as well as actions to build resilience for adapting to the impacts of climate change. Each country was to complete an NDC by 2020. The Paris Agreement works on a five-year cycle, where the NDCs will become increasingly ambitious over time.

The Paris Agreement recognizes that action is required by all while reaffirming that developed countries should take the lead. Support from Annex I countries to non-Annex I countries outlined in the Paris Agreement includes:

- **Finance:** Developed countries should provide additional financing to less developed and more vulnerable countries for both mitigation (emissions reductions) and adaptation. A significant difference from the KP is that the Paris Agreement also encourages voluntary financial contributions by non-Annex I Parties.
- **Technology:** Establishes a framework to accelerate technology development and transfer between parties.
- **Capacity-building:** Requests developed countries to increase support for climate-related capacity-building in developing countries.

The Paris Agreement also includes an enhanced transparency framework (ETF). Starting in 2024, countries will report on their actions and progress in mitigation and adaptation and will transparently report on support provided or received. The information gathered through this process will feed into a global stocktake every five years. The global stocktake will assess overall progress and inform countries as they set more ambitious plans in the next round.

🌐 **126**

35°C CAN BE FATAL

At a temperature of 35°C/95°F (wet bulb reading, 100 percent humidity), human beings cannot survive.

Indigenous Youth Represent Their Culture to Demand Action

In the months leading up to the 2021 United Nations Climate Change Conference (COP26) held in Glasgow, Scotland, a group of young activists from Panama's Indigenous Guna people came together to create a massive handmade "Mola" sail. They planned to bring it to Glasgow to bring attention to the effects of rising oceans on the Guna homelands. The traditional colorful handsewn mola cloth appliqué technique used for the sail is unique to the region.

The majority of the Guna people (about 33,000) live in Guna Yala, a province primarily located on the San Blas Islands in the Caribbean Sea. The islands are in danger of being rendered uninhabitable by sea-level rise in the next several decades. The Guna youth are active participants in Geo 2030, a 10-year leadership development campaign for global climate action. Planned in Panama in 2020 and launched globally in 2021, Geo 2030 is directed by a council of youth and elders representing a diversity of youth-led international and Indigenous organizations and companies.

The sail was handmade by 37 Guna artisans, and at 40 square meters is the largest Mola ever created. To ensure that the entire Guna people were represented in the project, the team of young people consulted with their elders.

Months in advance of the departure for Glasgow for COP26, the team sought approval to hang the Mola Sail somewhere inside the Blue Zone, the area reserved for official delegates, heads of state, and major corporate sponsors. However, nobody in any official capacity replied to their request.

The young activists then decided to unilaterally take action. In the early days of COP26, they staged demonstrations with the Mola Sail as the centerpiece. Then, halfway through COP26, they found an ideal location to hang the Mola Sail.

Without asking permission, they hung the sail near the main venue. They found people sympathetic to their cause with the right heavy equipment required and late one evening worked together to hang the huge Mola.

For the remainder of COP26, the Mola Sail was visible to all and served as a ceremonial gathering spot for Indigenous leaders. Media outlets like the BBC and newspapers including *The National* covered the Guna efforts.

"The Mola Sail represents the origin of my people's identity as

DISASTERS LOOMING

Children of this generation are three times more likely to face climate disasters than their parents did.

a Guna nation," says Agar Inklenia Tejada, a student of architecture and design associate with Geoversity Design who was a co-leader of the project. "It symbolizes our deep caring of our Mother the Earth with its honoring of the sky, sun, sea, earth and of all living beings. It's the sail that unites us and moves us forward to fight for the forests, rivers and oceans of our Mother."

"Our Geo 2030 action agenda is grounded in the hard work of preparing our ocean, riverside, and forest communities for the wrenching changes we have to make in the face of a rising sea, flooding rivers, landslides, incursions by loggers and ranchers, and parched forests on fire," says Iniquilipi Chiari, co-founder of the Guna Youth Congress and of Geoversity School of Biocultural Leadership. "We must continue growing stronger in our resolve, smarter in our organization and united in our action with our brothers and sisters of lands close by and far away."

🌐 **120**

What Are Cities Doing? (The C40)

Nearly 100 of the world's most influential cities, jointly called the C40, are working together to address climate change. These cities represent over 700 million people—more than a quarter of the global economy.

The mission of the C40 is to halve the greenhouse gas emissions of member cities within a decade, in line with the targets of the Paris Agreement. Developing climate action plans to describe concrete measures to cut emissions and increase urban resilience are part of the requirements.

C40 cities freely share advice on what works. The group's networks bring together officials from various cities working on similar climate actions. The city-to-city sharing of experience and best practices helps members cut costs, prevent mistakes, and build capacity.

The C40 also creates positive peer pressure to drive action. Once one city shows it can deliver on an ambitious goal, a new standard is set for all cities.

🌐 **125**

97
member cities
in the c40

25%
of the global economy
is made up of C40 cities

700+
million people make up
the members of the C40

The impact of city to city sharing

the number of C40 cities **restricting high-polluting vehicles** increased by over 700%

2009	→	2020
3 Cities		**23 Cities**

the number of C40 cities with a **cycle hire scheme** increased by over 600%

2009	→	2020
14 Cities		**86 Cities**

the number of C40 cities **incentivizing renewable electricity** increased by 650%

2009	→	2020
4 Cities		**26 Cities**

the number of C40 cities **investing to tackle flood risk** increased almost 1400%

2009	→	2020
4 Cities		**55+ Cities**

Schools and Solar Power

In 2016, K-12 schools in the US spent $8 billion on energy. This was 25 percent more than the spending level reported three years earlier.

In Europe, energy for schools makes up 70 percent of municipal energy expenses. In France, schools account for 30 percent of municipal building energy use.

At the other end of the spectrum, 291 million children—mostly in sub-Saharan Africa, South Asia, and Latin America—attend primary schools without electricity. Advances in these schools will either increase future CO_2 emissions over today's levels or require sustainable electrification.

Two main approaches have been taken to address both the energy costs and the environmental impact of energy consumption. Schools can reduce energy use and source the remaining energy from renewable sources.

The overlap between daylight and typical school hours means that the energy consumption profile of schools is compatible with solar power. During weekends and school holidays, energy can be stored, or where appropriate, exported to the grid, typically speeding up the payback of the system. One study reported, "If all the K-12 schools in the US were completely powered by the sun, it would eliminate CO_2 pollution equivalent to shutting down 18 coal-fired power plants."

A single program operating only in County Durham schools in the UK has saved £8.8 million, 11.2 tons of CO_2, and 202 GWh since it began in 2010.

Between 2014 and 2019, the number of US schools with solar installations increased by 80 percent to a total of 7,332 schools, making up 5.5 percent of K-12 schools. Schools also provide an opportunity to include sustainability in the curriculum of the leaders of tomorrow.

Many governments and organizations across Europe, the US, and Australia assist schools in starting their sustainability efforts by offering energy audits to identify savings. They then provide funding programs to offset the upfront installation costs. Depending on the region, there may be programs at national, state, and district or county levels.

Where upfront funding isn't available, there are other financing mechanisms such as grants and subsidy programs, lease arrangements, power purchase agreements, and more.

The largest financial gains typically come from funding the installation with school funds. These achieve payback within three to five years. For typical schools in OECD countries, there is a financial cost for every year of inaction.

⊕ **116**

Schools and solar power

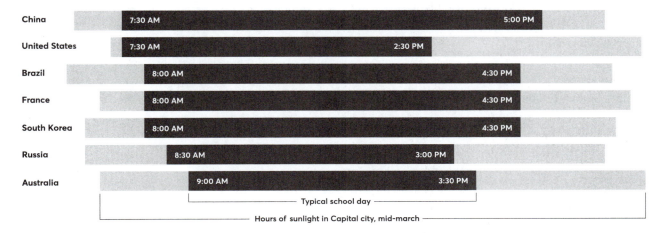

China	7:30 AM	5:00 PM
United States	7:30 AM	2:30 PM
Brazil	8:00 AM	4:30 PM
France	8:00 AM	4:30 PM
South Korea	8:00 AM	4:30 PM
Russia	8:30 AM	3:00 PM
Australia	9:00 AM	3:30 PM

Typical school day

Hours of sunlight in Capital city, mid-march

The State of Climate Change Litigation

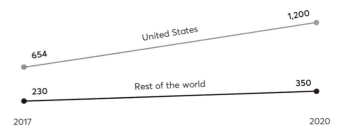

Active climate litigation cases

United States
654 — 1,200

Rest of the world
230 — 350

2017   2020

Why law matters

Around the world, activists are calling for more legislation and government action to address climate change. But in addition to new laws, litigation is a powerful tool for change. There are 1,843 active climate litigation cases currently in process.

The growth in climate litigation is noteworthy. In 2017, there were 884 cases brought forward in 24 countries. By the end of 2020, there were 1,550 cases active in 38 countries. About two–thirds of all cases are located in the US.

People can induce organizations to change behavior to limit greenhouse gases through climate-specific legislation and civil litigation to enforce that legislation. Climate litigation can also be applied to laws not written specifically for climate-related matters.

On December 20, 2019, the Dutch Supreme Court, the highest court in the Netherlands, upheld the previous decisions in the Urgenda Climate Case, finding that the Dutch government has obligations to urgently and significantly reduce emissions in line with its human rights obligations. This was a truly historic outcome.

Who is bringing forward cases, and against whom

Generally, defendants are either national governments or corporations, with governments still representing the majority. Plaintiffs come from all corners of society, including:

- activists
- individuals
- class action groups
- indigenous populations
- other governments (e.g., states bringing action against national governments)
- public and private financial institutions and regulators
- political parties

These groups bring litigation forward utilizing a wide range of legal theories, with varying degrees of success.

The theories on which cases are brought

The field of climate litigation is in what has been called an "exploration" phase, as groups seek to determine which bases for claims will prove most effective. The United Nations Environment Programme outlines several approaches:

- **Climate rights**: Litigants assert that insufficient action to mitigate climate change violates plaintiffs' international and constitutional rights to life, health, food, water, liberty, family life, and more.
- **Domestic enforcement**: Litigants assert that relevant laws and regulations are not being enforced.
- **Keeping fossil fuels in the ground**: These cases involve claims that corporations or government agencies involved in energy extraction projects have overlooked climate change implications in their environmental review processes.
- **Corporate liability and responsibility**: The plaintiffs attempt to attach causal responsibility for climate-related harm to defendants' actions.
- **Failure to adapt and the impacts of adaptation**: A range of cases seek to prove defendants have not exercised their duty to avoid harm.
- **Climate disclosures and greenwashing**: These suits, primarily against corporations, allege that defendants failed to properly disclose risk and other information about possible climate-related harm and thereby prevented effective decision-making by stakeholders.

The results

The results of climate litigation efforts are mixed. In many cases, courts have determined plaintiffs do not have proper standing to bring the case before the court. Courts have also used justiciability, where a court may not consider a plaintiff's claim on the grounds that it must be decided by another branch of government, such as an agency of the executive branch. While courts in the US have declined to hear many cases due to lack of standing, more cases in the developing world have been able to move ahead.

Beyond winning or losing, a case might also be evaluated based on impact. In The British Academy's COP26 briefing, "Climate Litigation as Climate Activism: What Works?," the authors lay out three categories of impact:

1. Cases that are won and have a reasonable potential to contribute to climate action, including halting specific industrial projects, as well as more sweeping cases such as Urgenda v. The Government of the Netherlands. In December 2019, the Dutch Supreme Court ruled the Dutch government must immediately reduce emissions in line with human rights obligations.

2. Cases that fail—often for reasons of justiciability—but garner significant public attention which can lead to further positive action. A notable example of such a case is Juliana v. The United States of America. This case brought by 21 youth in 2021 claimed the US has violated the youngest generation's constitutional rights to life, liberty, and property, and failed to protect public resources.

3. Very high profile cases, particularly against major energy concerns, that are unlikely to succeed in court but are intended to cause significant reputational impact and shift public narratives.

Litigation can take years to resolve, and the entire area of climate litigation is still relatively young. While results are uncertain, the rapidly growing volume of cases, actors, and strategies suggests litigation will remain part of the global response to climate change.

🌐 **121**

Key trends in climate change litigation

Claiming corporate liability and responsibility for climate harms

Seeking to keep fossil fuels in the ground

Addressing failures to adapt and the impacts of adaptation

Advocating for greater climate disclosures and an end to corporate greenwashing on the subject of climate change and the energy transition

Challenging domestic enforcements (and non-enforcements) of climate-related laws and policies

Increasing number of cases relying on fundamental and human rights enshrined in international law and national constitutions to compel climate action

The Positive Impact of Sustainability on Investor Returns

A growing body of research shows that companies that proactively address climate change provide better investment returns to their shareholders. There is money to be made in sustainability.

Climate change is frequently viewed as the biggest risk faced by investors. Actions to cut emissions and avoid the worst impacts of climate change are now seen as the best path to protect long-term investment value and returns.

Investors concerned about climate impacts often use a company's environmental, social, and governance (ESG) rating when making investment decisions. Companies committed to ESG initiatives (such as reducing carbon intensity, increasing renewable energy use, and recycling) are publishing measurable goals—and progress toward those goals—in periodic sustainability reports.

Increasingly, these reports are following the ESG standards established by the Global Reporting Initiative (GRI) and/or the United Nations Principles for Responsible Investment (PRI). The number of signatories has grown each year has ESG issues have received greater focus in the investment community.

ESG assets held by individual investors increased by 50 percent from 2018 to 2020, to $4.6 trillion from $3 trillion.

Global ESG assets are on track to exceed $53 trillion by 2025, representing more than one-third of the $140.5 trillion in projected total assets under management.

While Europe currently accounts for half of global ESG assets, the US may dominate the category starting in 2022. The next wave of growth could come from Asia, particularly Japan.

ESG–weighted investment portfolios are matching and sometimes exceeding the returns on traditional stocks indices.

As of November 29, 2021, investing in an ESG-weighted version of the S&P 500 would have provided a YTD return of 25.33 percent, three percent higher than the non-weighted S&P 500's return of 22.33 percent.

⊕ 131

Number of PRI signatories

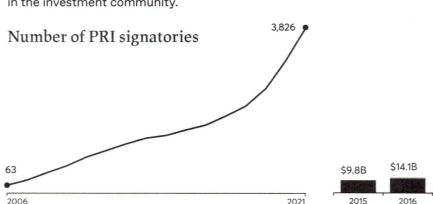

3,826

63

2006 — 2021

Yearly inflows into ESG mutual funds and ETFs ($B)

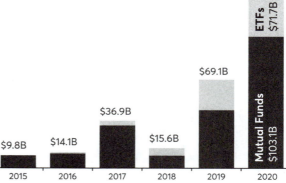

$174.8B

ETFs $71.7B

Mutual Funds $103.1B

$9.8B $14.1B $36.9B $15.6B $69.1B

2015 2016 2017 2018 2019 2020

Youth-Led Climate Litigation

A surprising portion of the nearly 2,000 climate litigation cases in process around the world today are initiated by young people, sometimes not even of voting age. These cases accuse national governments of failing to take action to prevent climate change, thereby violating rights to life as enshrined in national constitutions, UN Conventions, or European judicial institutions.

Four cases highlight one unique underlying principle—to redress harms that will *occur in the future.*

Juliana v. The United States of America: This constitutional climate case was filed in 2015 by 21 youths represented by Our Children's Trust. The plaintiffs allege the United States Government has "affirmatively" contributed to climate change and violated the youngest generation's constitutional rights to life, liberty, and property, as well as failed to protect essential public trust resources.

In February 2021, the Ninth Circuit Court affirmed a prior decision that the plaintiffs lacked the legal right to sue and urged the plaintiffs and government to work toward a settlement. After five months, the parties failed to reach resolution. As of December 2021, the courts are considering a request by the plaintiffs to file an amended complaint.

Saachi et al. v. Argentina, Brazil, France, Germany, and Turkey: Sixteen young people—including Greta Thunberg—from 12 nations brought claims against five nations under Article 5 of the Third Optional Protocol to the UN Committee on the Rights of the Child. This 1989 treaty is the most widely ratified treaty in history. Plaintiffs claim these nations have violated their rights by "exposing them to life-threatening dangers and harming their health and development."

In October 2021, the UN Committee ruled the claims inadmissible because "local remedies" had not been exhausted. However, members "accepted the claimant's arguments that States are legally responsible for the harmful effects of emissions originating in their territory on children outside their borders..."

Neubauer et al. v. Germany: This case was filed by a group of German youths in February 2020 against Germany's Federal Climate Protection Act. It claimed 55 percent target reduction in GHG by 2030 was insufficient to protect current youth and future generations. On April 29, 2021, the Federal Constitutional Court ruled in favor of the young people, claiming aspects of the German Basic Law represented a "legal norm that is intended to bind the political process in favor of ecological concerns, also with a view to the future generations that are particularly affected."

Sharma v. Minister for the Environment: In 2020, eight Australian youths filed a claim against the Australian Minister for the Environment. It claimed the approval of a coal mining project was a climate threat and breach of duty of care the government owes to future generations. In July of 2021, the Federal Court of Australia issued a series of rulings on the case. While declining to issue an injunction, the court found the Minister "has a duty to take reasonable care... to avoid causing personal injury or death to persons who were under 18 years of age and ordinarily resident in Australia at the time of the commencement of this proceeding arising from emissions of carbon dioxide into the Earth's atmosphere." The Executive Branch has challenged the validity of attributing any negative climate impact to the project and the Minister subsequently approved its construction. An appeal is currently being heard.

⊕ **134**

Youth-led climate change litigation is forcing courts and governments to confront... the devastating and disproportionate impact of climate change on our children and future generations.

— Marc Willers

Share of Global GHG Emissions Covered by Carbon Pricing Systems

Putting a price on carbon is widely recognized as *the* essential market-based tool to reduce global GHG emissions. It came to prominence in 2005 with the implementation of the European Emissions Trading System. From covering 5.3 percent of global emissions at that time, carbon pricing initiatives rose gradually to cover 21.5 percent of global emissions in 2021. In 2019, governments raised approximately US $45 billion in revenues with carbon pricing initiatives.

The largest annual increase occurred in 2021, with the implementation of China's emissions trading system, which covers about 7.4 percent of global emissions.

⊕ **838**

PLASTIC BRICKS

A new technology called ByBlocks converts discarded plastic into materials that can be used instead of concrete. Because the plastic is compressed rather than melted or transformed, it can incorporate a wide variety of plastic, avoiding incineration or landfills.

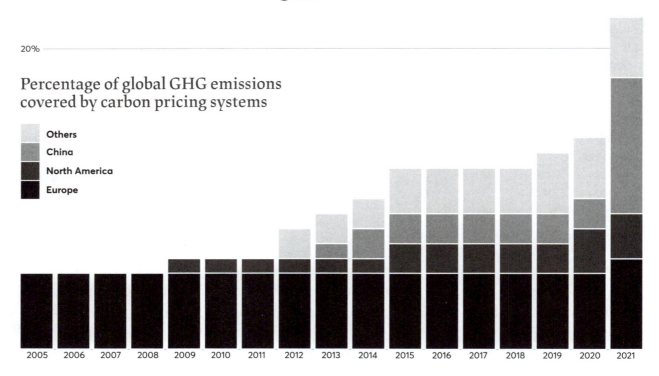

Percentage of global GHG emissions covered by carbon pricing systems

- Others
- China
- North America
- Europe

20%

2005 2006 2007 2008 2009 2010 2011 2012 2013 2014 2015 2016 2017 2018 2019 2020 2021

Role of Finance

Growing a business requires access to capital—and the more capital-intensive the business, the greater will be its need for financing.

Energy companies tend to be capital-intensive: The costs of drilling and operating an oil field or building wind farms and energy plants quickly add up.

Financiers focus on lending to firms that are likely to pay off their debts, and fossil fuel companies have traditionally been considered a safe bet. As such, the major fossil fuel companies have enjoyed easy access to capital, chiefly in the form of loans.

The largest banks have continued to finance fossil fuel companies since the 2015 Paris Accords. In fact, lending to those companies has continued to increase by about five percent per year—$824 billion in total in 2019, which was $43 billion more than in 2018. In total, Bloomberg estimates that fossil fuel companies have secured approximately $3.6 trillion worth of new loans since 2015. Banks have collected more than $16.5 billion in fees on the trillions lent to fossil fuel companies since 2015.

For alternative energy companies and projects to succeed, they, too, need access to capital. And although "green" debt issuance has grown substantially since 2015, it only amounts to one-third of fossil–fuel–related financing over the same period.

In 2021, green companies attracted more financing than fossil fuel companies for the first time. However, primarily due to COVID–related disruptions, they attracted less total financing than they did in 2020.

If post-2015 energy financing trends continue, it is unlikely the world will be able to transition away from fossil fuels quickly enough to meet the 1.5°C/2.7°F goal agreed upon at Paris.

According to an analysis by the Rainforest Action Network, continuing current levels of investment will keep fossil fuels' share of the global energy mix at over 75 percent in 2030. Yet, models by McKinsey & Company show that fossil fuels must provide no more than half the world's energy by 2030—and none by 2050—to keep on track for less than 1.5°C/2.7°F of warming.

⊕ **836**

Percent of energy sector loan fees for 'green' projects 2016 through mid 2021

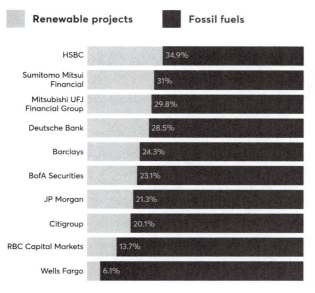

Renewable projects   Fossil fuels

- HSBC — 34.9%
- Sumitomo Mitsui Financial — 31%
- Mitsubishi UFJ Financial Group — 29.8%
- Deutsche Bank — 28.5%
- Barclays — 24.3%
- BofA Securities — 23.1%
- JP Morgan — 21.3%
- Citigroup — 20.1%
- RBC Capital Markets — 13.7%
- Wells Fargo — 6.1%

Annual debt issue

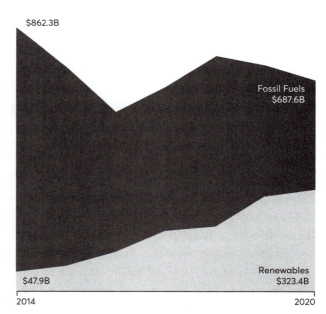

$862.3B

Fossil Fuels $687.6B

Renewables $323.4B

$47.9B

2014   2020

ESG Reporting Frameworks

Environmental, social, and governance (ESG) data are used to prove a company's effectiveness in reducing carbon emissions. Companies use this data to measure their impact, and investors use it to determine how climate risks affect their investments.

Companies often use several frameworks when making ESG disclosures. They also choose which part of each framework to use. These factors are raising questions about accuracy and reliability. In a 2020 report, ESG data quality topped a list of investor concerns.

There are currently six major global reporting frameworks and several dozen less-common options. Some frameworks have started to consolidate, leading to the newly formed International Sustainability Standards Board (ISSB).

The most common ESG frameworks are:

1. Global Reporting Initiative (GRI)
Year created: 1997

Usage: 72 percent of 250 of the world's largest companies and 67 percent of the 100 largest firms in 52 countries

Description: GRI refers to itself as "the world's most widely used standards for sustainability reporting." It publishes a set of standards as well as formats for required and recommended disclosures. Standards are set on materials, energy, water, biodiversity, emissions, pollution, waste, and supplier impact. These standards highlight a company's impact on the world.

2. Carbon Disclosure Project (CDP)
Year created: 2000

Usage: Over 13,000 companies and 1,100 cities, states, and regions and nearly 600 investors with over $110 trillion in assets under management

Description: CDP "supports thousands of companies, cities, states, and regions to measure and manage their risks and opportunities on climate change, water security, and deforestation." It focuses on quantitative environmental impact data, using an independent approach to review reports and assign letter grades. In 2021, over 270 companies received an A rating for climate change, forests, or water security.

3. Principles for Responsible Investment (PRI)
Year created: 2006

Usage: Over 4,500 signatories, with 75 percent of them investment managers

Description: PRI says it is the "world's leading proponent of responsible investment." It provides six voluntary principles, each broken down into multiple possible actions. Signatories submit their data annually using an online reporting tool. PRI publishes each signatory's reports and favors independent validation of data. The number of signatories has grown each year has ESG issues have received greater focus in the investment community.

4. Sustainability Accounting Standards Board (SASB)
Year created: 2011

Usage: 1,271 active users globally and supported by 258 institutional investors representing $76 trillion in assets under management across 23 countries

Description: SASB's accounting-focused standards "enable businesses around the world to identify, manage and communicate financially–material sustainability information to their investors." They emphasize reporting information that is financially material to companies and investors rather than environmental impact. There are specific standards for 77 industries because each industry has different carbon and other ESG impacts. A new organization called the Value Reporting Foundation (VRF) now maintains SASB standards.

5. Task Force for Climate-related Financial Disclosure (TCFD)
Year created: 2015

Usage: 89 jurisdictions and over 2,600 organizations, including financial institutions with $194 trillion in assets under management and non–financial companies with $25 trillion in market capitalization

Description: TCFD "develop[s] recommendations for more effective climate-related disclosures that could promote more informed investment, credit, and insurance underwriting decisions." It focuses on carbon and climate risks that create investment exposures in the global financial system. TCFD asks organizations to disclose in four main areas: the role of boards of directors and leaders, strategic scenario analysis and planning, the ability to assess and manage climate risks, and current metrics and future targets. Some guidance covers all sectors, while other principles cover specific industries.

6. United Nations Sustainable Development Goals (UN SDGs)

Year created: 2015
Usage: Over 15,000 companies of all sizes
Description: The UN SDGs are general principles for humanity "to end extreme poverty, fight inequality and injustice, and protect our planet" rather than a specific measurement framework. There are 17 Goals with 231 unique indicators such as "Total greenhouse gas emissions per year." Companies using the SDGs create a Communication on Progress (CoP), an overall report that includes non–financial data and information about what they have done to help address SDG principles. It might also include measurement using other frameworks. The intended audience includes policymakers, community stakeholders, the general public, and investors.

🌐 **124**

Global Companies Committing to Science-Based Net-Zero Targets

Companies setting net–zero targets by 2050
As of November 2021, 1,045 companies are setting near-term targets aligned with a 1.5°C/2.7°F scenario. This comes with a commitment for actions that lead to net zero by 2050.

Together, these companies:
- Span 60 countries, 53 sectors, and more than 32 million employees
- Amount to 23 trillion dollars in market value, equivalent to the size of the US economy

Their commitment would result in result in 262 million tons of emissions being cut by 2030 (equivalent to the annual emissions of Spain).

Science behind the targets
Net zero means the amount of greenhouse gas emissions released into the atmosphere equals the amount of greenhouse gas emissions removed from the atmosphere. Although it's sometimes used interchangeably with *carbon neutral*, there is an important difference. Companies that claim carbon-neutral status often rely heavily on purchasing carbon offsets, instead of actually reducing their own emissions.

Science-Based Targets aim to shift companies firmly towards emissions reduction. They provide a global standard for companies to set science–based net–zero targets that are aligned with the Paris Agreement goal of limiting the rise in global temperature to 1.5°C/2.7°F.

The approach is based on fairly allocating a portion of the global carbon budget to a company based on its size and its activities. It considers not only direct production but also the supply chain and life cycle of company products. A pathway for reducing emissions to net zero is then plotted, based on the latest scientific scenarios developed by the Intergovernmental Panel on Climate Change and the International Energy Agency.

In addition to a commitment to net zero by 2050, concrete emissions reduction targets are set for the next 5-10 years, with progress tracked annually. This is to ensure emissions are actually reduced by 2030, when it is most needed.

🌐 **112**

Where You Bank Makes a Difference

Trillions of dollars make a difference. The deposits in checking and savings accounts support investments by commercial banks in the fossil fuel economy.

Specifically, about 95 percent of households in the United States have one or more checking or savings accounts in commercial banks. These 124,000,000 households have an average of more than $40,000 in their accounts, which means the banks are holding over **$5 trillion of American consumers' money**. This is only in checking and savings accounts: Banks hold other financial assets of individuals and families as well.

Commercial banks invest money in three ways: making loans to businesses and individuals, buying securities either as individual stocks or funds, and simply holding money in places where it can earn interest at greater rates than it pays to depositors.

How banks decide to make those loans or investments is generally driven by the need to generate significant financial returns.

Commercial banking is increasingly concentrated in the hands of just a few institutions: About 45 percent of all deposits are in four banks: JP Morgan Chase, Citibank, Wells Fargo, and Bank of America.

Overall, from 2016-2019, the top 35 banks globally (including American banks) have invested over $2.7 trillion in fossil fuel companies. The top four alone invested $811 billion in the fossil fuel economy during this time. Banking support for fossil fuels has *increased* since the signing of the Paris Accords.

However, an increasing number of commercial banks of all sizes are establishing policies that limit fossil fuel investments and lay out a transition plan for the future. This creates an opportunity for individuals to have an impact on a bank's investment choices without foregoing the benefits of having bank accounts.

Websites such as Mighty or the Global Alliance for Banking On Values identify which banks have sustainability and/or fossil fuel-free lending and investing policies.

An individual with $40,000 in a checking or savings account can have an impact on the lending and investing practices of a trillion-dollar bank.

"Yes, the planet got destroyed. But for a beautiful moment in time we created a lot of value for shareholders."

1. Withdrawing funds from a bank reduces the amount of funds they can access for further investment in fossil fuels.

2. Explaining to the bank why the funds are being moved—and then sharing that reason on social media—creates social pressure that can prompt others to do the same and create a cascade effect.

It is a powerful signal akin to shareholder activism or divestiture movements. The four largest banks have entire teams of executives focused on their market share. They're listening.

More details about the process in general and for each bank can be found at Banks.org or Chime.com.

🌐 **133**

> *It is not an option but rather an obligation to speak out, all of us together, to demand changes.*
>
> — Betty Vasquez

Top 20 banks by fossil fuel financing, 2016–2019

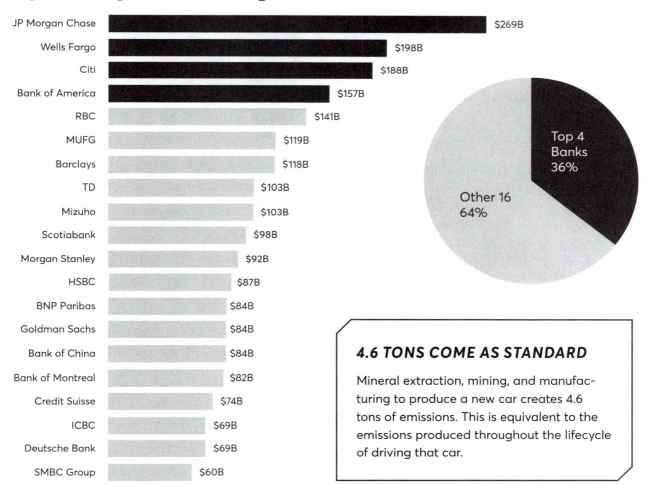

JP Morgan Chase	$269B
Wells Fargo	$198B
Citi	$188B
Bank of America	$157B
RBC	$141B
MUFG	$119B
Barclays	$118B
TD	$103B
Mizuho	$103B
Scotiabank	$98B
Morgan Stanley	$92B
HSBC	$87B
BNP Paribas	$84B
Goldman Sachs	$84B
Bank of China	$84B
Bank of Montreal	$82B
Credit Suisse	$74B
ICBC	$69B
Deutsche Bank	$69B
SMBC Group	$60B

Top 4 Banks 36%

Other 16 64%

4.6 TONS COME AS STANDARD

Mineral extraction, mining, and manufacturing to produce a new car creates 4.6 tons of emissions. This is equivalent to the emissions produced throughout the lifecycle of driving that car.

The 20 Largest Fossil Fuel Producers

Fossil fuel companies are some of the largest and most profitable in the world. The product they create is responsible for a significant percentage of carbon emissions. Their lobbying and promotional efforts are designed to protect the value of their assets in the ground.

- 60 percent are owned by governments
- 40 percent are investor-owned

THE 3rd LARGEST OIL RESERVE

Canada ranks third in oil reserves worldwide. Their untapped oil is currently 188 times their annual consumption.

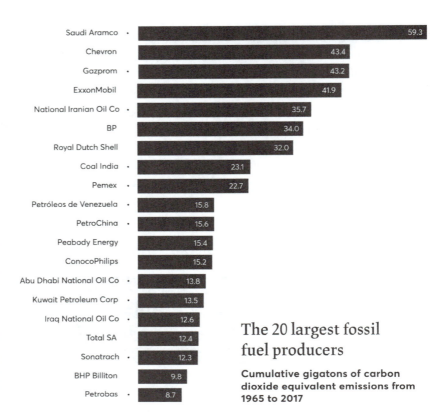

Saudi Aramco • 59.3
Chevron 43.4
Gazprom • 43.2
ExxonMobil 41.9
National Iranian Oil Co • 35.7
BP 34.0
Royal Dutch Shell 32.0
Coal India • 23.1
Pemex • 22.7
Petróleos de Venezuela • 15.8
PetroChina • 15.6
Peabody Energy 15.4
ConocoPhilips 15.2
Abu Dhabi National Oil Co • 13.8
Kuwait Petroleum Corp • 13.5
Iraq National Oil Co • 12.6
Total SA 12.4
Sonatrach • 12.3
BHP Billiton 9.8
Petrobas • 8.7

The 20 largest fossil fuel producers

Cumulative gigatons of carbon dioxide equivalent emissions from 1965 to 2017

Here is the contact information for the public companies on the list:

BP
1 St. James's Square London UK SW1Y 4PD
+1-800-333-3991 (US)

BHP Billiton
171 Collins Street Melbourne Victoria 3000 Australia
+61-3-1300-55-47-57

Chevron (NYSE: CVX)
6001 Bollinger Canyon Road San Ramon CA 94583 USA
+1-925-842-1000

ConcocoPhilips
925 N. Eldridge Parkway Houston Texas 77079 USA
PO Box 2197 Houston TX 77252-2197 USA
+1-281-293-1000

ExxonMobil (NYSE: XOM)
5959 Las Colinas Boulevard Irving Texas 75039-2298 USA
+1-972-940-6000

Peabody Energy
Peabody Plaza 701 Market St. St. Louis MO 63101-1826 USA
+1-314-342-3400

Total SA
Charl Bosch Street Sasolburg South Africa 9570
+27-11-283-4900

Royal Dutch Shell Pc
Carel van Bylandtlaan 16 2596 HR The Hague The Netherlands
PO Box 162 2501 AN The Hague The Netherlands
+31-70-377-911

⊕ 114

10 Publishers Promoting Climate Change Denial Content Online

Where does climate change denial live online? The Center For Countering Digital Hate (CCDH) sampled nearly 7,000 climate denial posts on Facebook. The publishers listed below were responsible for 69 percent of all interactions with climate denial content on social media. Almost 99 percent of those user interactions were with unlabeled posts.

Collectively, this group has up to 186 million followers on mainstream social media platforms, and their websites received nearly 1.1 billion visits in the last six months of 2021.

🌐 345

THE ONE PERCENT'S EMISSIONS

The emissions created by the richest one percent of the global population account for more than double the emissions of the poorest 50 percent.

Percentage of interactions with climate denial content

PUBLISHER	PERCENTAGE
Breitbart	17.1%
The Western Journal	15.6%
Newsmax	9.9%
Townhall Media	6.5%
Media Research Center	6.1%
The Washington Times	6.0%
The Federalist Papers	2.4%
The Daily Wire	2.0%
Russia Today	1.8%
Patriot Post	1.6%

Just start by doing something, anything, and then talk about it! Talk about how it matters to your family, your home, your city, the activity that you love. Connect the dots to your heart so you don't see climate change as a separate bucket but rather as a hole in the bucket of every other thing that you already care about in your life.

Talk about what positive, constructive actions look like that you can engage in individually, as a family, as an organization, a school, a place of work. Add your hand to that giant boulder. Get it rolling down the hill just a little faster.

— Dr. Katharine Hayoe

Oil Subsidies

The Industrial Revolution kicked off the need for reliable energy sources for factories, and countries with a solid industrial base grew faster and became more powerful as a result. This increased demand for fuel.

To have a consistent supply of fossil fuels, governments subsidize them by giving financial aid (such as direct cash payments or tax breaks) to producers to lower production costs or increase the price received from fossil fuels. Governments can also lower the price paid by consumers.

Reducing the cost of production or increasing the price of oil when its price is too low allows producers to remain profitable. If the price of oil happens to be too high for consumers, governments can subsidize it by directly giving cash to consumers or through other indirect measures such as tax exemptions.

Globally, $447 billion in subsidies are provided for fossil fuels and only $128 billion in subsidies for renewable energies. However, total subsidies for fossil fuels have been declining.

Don't take our word for it

Visit **thecarbonalmanac.org/123** to check out this article's sources, relevant links, and updates.

Dig deep and share what you learn.

Cloud Seeding

Cloud seeding artificially increases condensation in clouds in order to produce more rain or snow. It is also used as a method for hail suppression.

Cloud seeding technology, a form of weather modification, has been around for more than half a century, and over fifty countries use it to alter weather. Funding for cloud seeding comes from various sources including insurance companies trying to mitigate hail damage, federal and local governments attempting to increase water in reservoirs, and ski resorts bulking up snow production. Hydroelectric companies use cloud seeding as well because more snow means more runoff in the spring, which in turn means more water for electricity.

There are two forms of cloud seeding: intentional and unintentional. Intentional cloud seeding is the active and planned injection of compounds into existing clouds (note that seeding cannot create clouds). Unintentional seeding is the natural seeding of clouds with biological "dirt" such as pollen or the unnatural and detrimental seeding of clouds with human-produced pollution.

How cloud seeding works

Cloud seeding modifies a cloud's structure by adding small, ice-like particles—typically silver iodide particles—to clouds.

These particles act as additional condensation nuclei. Unattached supercooled water vapor molecules in the clouds condense around these particles. The condensed water vapor droplets group together and the process continues until droplets are large enough to fall as rain.

There are two ways of adding particles to clouds:

1. large cannons that shoot particles into the sky
2. airplanes that drop particles from above

Environmental impact

The impacts of cloud seeding are challenging to measure, with research suggesting it might create a 10-15 percent increase in rainfall. How much rain or snowfall would have occurred without the intervention is an open question. Another unknown is whether manipulating weather in one section of clouds ultimately affects the natural rain or snowfall in a nearby area.

Air pollution has a negative impact on precipitation since tainted air creates clouds with reduced droplet size. All clouds need dirt to form, and aerosols like sea salt, dust, and pollen create large particles ultimately resulting in large raindrops.

Longitudinal research from Israel has shown that air pollution creates an environment that impedes the positive effects of cloud seeding; the areas that most need increased rain and snowfall are least likely to create conditions for successful seeding. In addition, silver iodide—the material used in cloud seeding—is toxic to aquatic life, so precipitation from seeded clouds could potentially harm the environment.

🌐 **117**

Wealth and Greenhouse Gases

Change in CO_2 emissions and GDP per capita over time (1990 as baseline)

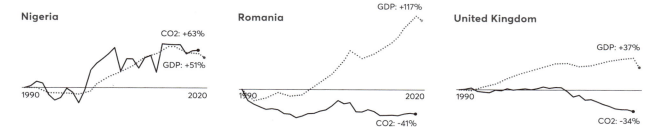

Is it possible to increase wealth while decreasing carbon emissions?

GHGs are strongly tied to Gross Domestic Product (GDP), a quantitative measure of human activity. The (apparently) cheap price of fuel has traditionally been a boon to productivity, enabling countries that burn more to profit. This means that countries that consume or produce more tend to emit more.

Despite this, between 2008 and 2018, many countries that saw a decline in their CO_2 emissions simultaneously grew their economy. Most achieved this thanks to a combination of increasing reliance on renewables and a declining use of coal-fired power plants. Attention to sustainability can lead to energy efficiency gains, and a gradual shift towards service industries (financial services, hospitality, IT, etc.).

🌐 **132**

GDP per capita vs yearly CO_2 emissions per capita

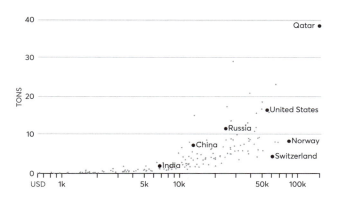

YOU CAN MAKE A DIFFERENCE

Visit **www.thecarbonalmanac.org** and sign up for **The Daily Difference**, a free email that will connect you with our community. Every day, you will join thousands of other people connecting around specific actions and issues that will add up to a significant impact.

The average carbon footprint of a US citizen (16 tons) is four times bigger than the global average of 4 tons.

Individual Carbon Footprint and Collective Action

It's not possible to solve the climate crisis solely by voluntarily scaling back individual activities. Still, awareness of those activities may help lower a person's impact, deepen understanding of the urgency of the issue, and might provoke them to take systemic action. Individuals can begin in their households, move on to the community, and then work across industries, nations, and the planet. Every step contributes to widespread change.

For many people, the most significant direct contribution to greenhouse gas emissions is driving a gasoline-powered vehicle. That's because every time a driver fills up the car, they're personally turning million-year-old carbon into CO_2.

Indirect contributors include things like building a house, purchasing a new pair of running shoes, or eating a grapefruit. Each involves manufacturing, shipping, storage, and other activities that rely on carbon to be accomplished. Each part of the supply chain affects the environment.

Beyond the propaganda

It's been widely reported that British Petroleum (BP) worked with Ogilvy to promote the term "carbon footprint" as a way to distract consumers from the significant impact industry was having on the environment. If people saw themselves as contributors to the problem, it would take the pressure off the systems that were profiting from investing in the industrialized combustion of carbon.

In the decades since the term caught on, many organizations and websites have made it easier for consumers to measure their personal impact and, if they desire, to tax themselves to address some of the damage.

But the real utility of this calculation is to remind people that systems approaches are the only way to solve a systems problem.

Carbon footprint calculation

Carbon calculators measure the carbon footprint of a household by factoring in how many people live at the location, the size of the home, transportation to work and other places nearby, how often air and other public transportation is utilized, and eating and shopping practices. The carbon footprint estimate is expressed by the tons of carbon dioxide emitted per year.

Once a carbon footprint is known, individuals can make different choices. They can reduce impact through lifestyle, e.g., reducing use of air conditioning, riding a bike to work, or taking a vacation near home rather than flying to another part of the world.

Carbon credits

Individuals may choose to offset their existing carbon footprint by funding an equivalent carbon dioxide saving elsewhere, often in developing countries. Common carbon offsets include funding reforestation projects that help to reverse the effects of greenhouse gases. Additionally, some airlines allow passengers to pay extra when booking a ticket to offset their portion of carbon generated by the flight.

🌐 **119**

All truth passes through three stages: First, it is ridiculed. Second, it is violently opposed. Third, it is accepted as self-evident.

— Arthur Schopenhauer

Talking About Climate Change

Communicating about climate change is an essential step in causing change to happen. There's still a long journey ahead: In 2021, only 31 percent of people surveyed said that climate change was a "top concern."

But changing people's minds is generally no easy task: Sharing facts and data shifts the behavior of people with differing worldviews only three percent of the time. However, engaging in a discussion that encourages self-generated reasons for change spikes the success rate to 37 percent.

Motivational interviewing has been studied as a technique for peer-to-peer opinion change. This is done by locating, connecting, and strengthening a person's own motivation and commitment toward impactful positive change.

The four rules of motivational interviewing are:

1. **Open-ended questions.** Get curious and avoid questions that can be answered with a simple "yes" or "no." For example: "I'm really curious to hear your thoughts and perspective on climate change. How could you see it affecting your grandkids?"

2. **Affirmations.** Acknowledging and emphasizing strengths can open a person up to a genuine exchange of ideas. Affirmations need to be genuine to be effective.

3. **Reflective listening.** Letting people speak freely builds a foundation for reflective listening. After someone has shared, an effective response is to neutrally mirror back what was heard. This confirms mutual understanding and helps people feel heard. The goal is to allow people to feel ownership over their decisions and actions. Reflective listening statements can take three forms, each corresponding with a deeper level of rapport: repeating or mirroring, paraphrasing, and reflection of feeling.

4. **Summaries.** By recapping what someone said, the listener creates space for the speaker to correct misunderstandings and address information gaps. This opens the door for more engagement.

🌐 **127**

Saint Kateri Habitats

Saint Kateri Habitats are small designated areas meant to nurture and restore natural environments. The initiative is a program of the Catholic faith-based Saint Kateri Conservation Center, founded in 2000.

A designated Saint Kateri Habitat can be any kind of natural landscape, such as rooftop container gardens, personal yards, community gardens, parks, meadows or farms. While being part of an ecosystem, they also serve as a sacred space.

As outlined by the Saint Kateri Conservation Center, such habitats always contain a religious artifact or icon plus at least two of the following features:

- Food, water, cover, and space for wildlife, including habitat for pollinators and other terrestrial and aquatic organisms
- Native trees, shrubs, herbaceous plants, and ecosystems
- Vegetable gardens, flower gardens, community-supported gardens, indoor gardens, and farms
- Ecosystem services, clean air and water, and carbon storage for climate regulation
- Renewable energy and sustainable practices of gardening, landscaping, and farming
- Sacred spaces for worship, prayer, and contemplation, including Mary gardens, prayer gardens, and rosary gardens

With 190 habitats on five continents, the Saint Kateri Habitats provide a spiritual approach to addressing climate change and biodiversity loss. Offering a place for meditation and reflection as well as pollination and propagation, the habitats bring nature to spaces in the human world and spiritual connection for the humans who enjoy them.

🌐 **130**

DEAD & COMPANY

It's been 57 years since the Grateful Dead was founded and the music never stopped. The Dead & Company 2021 tour was climate positive, eliminating 5X more greenhouse gas emissions than the tour created including fan travel to and from shows. The band used to play for silver, now they play for life.

Top Donors in Climate Philanthropy 2020–2021

In 2021, there was a shift in the landscape of philanthropic giving. Climate change mitigation received the largest-ever private funding commitment of $5 billion. In addition, new donors joined a pledge made in 2018 at the Global Climate Action Summit, promising an additional $6 billion by 2025.

2020 saw a 14 percent increase in climate change related giving in just one year, and climate change mitigation is estimated to represent two percent of total global philanthropic giving.

> ## *Destroying rainforest for economic gain is like burning a Renaissance painting to cook a meal.*
>
> — E. O. Wilson

The topic of philanthropic giving is complex, and many aspects are difficult to quantify. Some donors keep their information private. Specific major pledges are meant to be disbursed over an agreed-upon period of time, which can span a few decades. There may be a time lag between the funding commitment and when the data are made available to the public. The US IRS allows taxpayers to deduct charitable contributions up to 50 percent of their adjusted gross income, causing some to question the giver's intent.

With these complexities in mind, here's a list of the top donor commitments in climate philanthropy for 2020-2021:

TOP DONOR COMMITMENTS 2020-2021

- Arcadia
- Bezos Earth Fund
- Bloomberg Philanthropies
- Breakthrough Energy Ventures (fund backed by Bill Gates)
- Chan Zuckerberg Initiative
- Christensen Fund
- David and Lucile Packard Foundation
- Ford Foundation
- Good Energies Foundations
- Gordon and Betty Moore Foundation
- IKEA Foundation
- John D. and Catherine T. MacArthur Foundation
- Laurene Powell Jobs
- Nia Tero
- Oak Foundation
- Rainforest Trust
- Re:wild
- Rob and Melani Walton Foundation
- Rockefeller Foundation
- Sobrato Philanthropies
- Stewart and Lynda Resnick
- Tesla and the Musk Foundation
- William and Flora Hewlett Foundation
- Wyss Foundation

🌐 **839**

Leading the Way

People and organizations
guiding us to a better future

30 Leading Climate Scientists

The news agency Reuters compiled a list of the top 1,000 most influential climate scientists, based on three factors:

- The number of climate-related academic papers published by each scientist, measuring their productivity in the field;
- How distinguished each scientist is within their field, measured by the ratio of each paper's citations to the average citations received by paper in that specific field;
- Each scientist's impact on the non-academic world, measured by an aggregate score of mentions on social media, mainstream media, public policy papers, and sites like Wikipedia.

This table lists Reuters' top 30 scientists.

⊕ **138**

RANK	SCIENTIST	COUNTRY	TOP THREE INTERESTS	TOP THREE FIELDS OF RESEARCH
1	Keywan Riahi International Institute for Applied Systems Analysis	Austria	Energy systems Integrated assessment model Policy	Economics Applied Economics Environmental Science
2	Anthony A. Leiserowitz Yale University	United States	Climate Change Perception Policy	Studies in Human Society Psychology and Cognitive Sciences Psychology
3	Pierre Friedlingstein University of Exeter	United Kingdom	Climate Change Carbon Cycle Climate	Earth Sciences Biological Sciences Atmospheric Sciences
4	Detlef Peter Van Vuuren Utrecht University	Netherlands	Integrated Assessment Model Climate Change GHG Emissions	Economics Applied Economics Environmental Sciences
5	James E Hansen Utrecht University	United States	Climate Change Warming Climate	Earth Sciences Physical Geography and Env. Geoscience Atmospheric Sciences
6	Petr Havlfk International Institute for Applied Systems Analysis	Austria	Climate Change Gas Emissions GHG Emissions	Environmental Sciences Economics Applied Economics
7	Edward Wile Maibach George Mason University	United States	Climate Change Perception Beliefs	Medical & Health Sciences Public Health & Health Services Psychology & Cognitive Sciences
8	Josep G Canadell Commonwealth Scientific and Industrial Research Organisation	Australia	Climate Change Carbon Cycle Sink	Biological Sciences Environmental Sciences Earth Sciences
9	Sonia Isabelle Seneviratne ETH Zurich	Switzerland	Soil Moisture Moisture Climate	Earth Sciences Physical Geography and Env. Geoscience Atmospheric Sciences
10	Mario Herrero Commonwealth Scientific and Industrial Research Organisation	Australia	Climate Change Livestock Production	Agricultural & Veterinary Sciences Environmental Sciences Environmental Science and Management
11	David B Lobell Stanford University	United States	Yield Climate Change Crop Yield	Agriculture & Veterinary Sciences Crop & Pasture Production Biological Sciences

RANK	SCIENTIST	COUNTRY	TOP THREE INTERESTS	TOP THREE FIELDS OF RESEARCH
12	Ken Caldeira Carnegie Institution for Science's Department of Global Ecology	United States	Ecosystems Ocean Species	Biological Sciences Ecology Earth Sciences
13	Kevin E Trenberth National Center for Atmospheric Research	United States	Ocean Precipitation Variability	Earth Sciences Atmospheric Sciences Oceanography
14	Stephen A Sitch University of Exeter	United Kingdom	Climate Change Vegetation Model Climate	Biological Sciences Earth Sciences Ecology
15	Glen P Peters Center for International Climate and Environmental Research	Norway	CO_2 Emissions Climate Change Budget	Earth Sciences Economics Applied Economics
16	Ove Hoegh-Guldberg University of Queensland	Australia	Reefs Corals Coral Reefs	Biological Sciences Ecology Environmental Sciences
17	Richard Arthur Betts Met Office	United Kingdom	Climate Change Climate Warming	Earth Sciences Atmospheric Sciences Physical Geography and Env. Geoscience
18	Michael G Oppenheimer Princeton University	United States	Climate Change Ice Sheet Sea Level Rise	Earth Sciences Physical Geography and Env. Geoscience Environmental Sciences
19	William Neil Adger University of Exeter	United Kingdom	Climate Change Policy Livelihoods	Studies in Human Society Environmental Sciences Environmental Science and Management
20	William Wai Lung University of British Columbia	Canada	Climate Change Fisheries Ecosystems	Biological Sciences Ecology Environmental Sciences
21	Peter M Cox University of Exeter	United Kingdom	Climate Climate Change Warming	Earth Sciences Atmospheric Sciences Biological Sciences
22	Christopher B Field Stanford University	United States	Ecosystems Elevated CO_2 Species	Biological Sciences Plant Biology Ecology
23	Shinichiro Fujimori Kyoto University	Japan	Climate Change Computable General Equilibrium Model General Equilibrium Model	Economics Applied Economics Engineering
24	Elmar Kriegler Potsdam Institute for Climate Impact Research	Germany	Climate Policy Economics Integrated Assessment Model	Economics Applied Economics Environmental Sciences
25	Yadvinder Singh Malhi University of Oxford	United Kingdom	Forest Tropical Forests Ecosystems	Biological Sciences Ecology Environmental Sciences
26	Carlos Manuel Duarte King Abdullah University	Saudi Arabia	Ocean Climate Change CO_2	Earth Sciences Oceanography Biological Sciences
27	Chris D Thomas University of York	United Kingdom	Species Climate Change Butterflies	Biological Sciences Environmental Sciences Ecology
28	Stephane Hallegatte World Bank	United States	Climate Change Natural Disasters Policy	Economics Applied Economics Earth Sciences
29	Andy P Haines London School of Hygiene & Tropical Medicine	United Kingdom	Health Outcomes Heart Disease Risk Factors	Medical and Health Sciences Public Health and Health Services Clinical Sciences
30	Michael Obersteiner International Institute for Applied Systems Analysis	Austria	Prices Climate Change Land	Environmental Sciences Economics Applied Economics

Avoiding climate breakdown will require cathedral thinking. We must lay the foundation while we may not know exactly how to build the ceiling.

— Greta Thunberg

When a flower doesn't bloom, you fix the environment in which it grows, not the flower. — Alexander Den Heijer

Countries Leading Climate Change Action

The Climate Change Performance Index (CCPI) tracks countries' climate responsibility performance. The CCPI reports on the climate action of 57 countries, together accounting for over 90 percent of global GHG emissions. The report is a compilation of the work of 400 experts who evaluate each country's national and international climate policies.

In the Top 10 countries report for 2022, several of the top categories are blank because no country ranked 'very high' in implementing sufficient measures to alleviate climate change in these areas.

⊕ **145**

The factors in the scoring are:

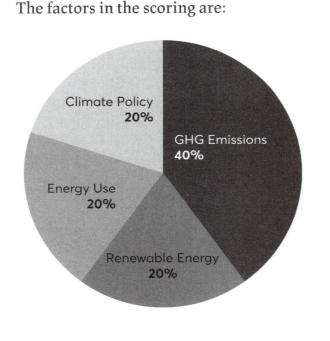

Climate Policy 20%
GHG Emissions 40%
Energy Use 20%
Renewable Energy 20%

RANK	COUNTRY OVERALL	RENEWABLE ENERGY	CLIMATE POLICY
1	–	–	–
2	–	–	–
3	–	Norway 19.21	–
4	Denmark 76.67	Denmark 14.93	Luxembourg 18.11
5	Sweden 74.22	Sweden 14.72	Denmark 17.87
6	Norway 73.29	Finland 4.04	Morroco 17.23
7	UK 73.09	Latvia 13.79	Netherlands 16.53
8	Morocco 71.60	New Zealand 13.05	Lithuania 16.48
9	Chile 69.51	Brazil 12.70	Portugal 16.27
10	India 69.20	Chile 12.62	France 16.06

African Reforestation Initiative

Launched in 2015, AFR100 (the African Forest Landscape Restoration Initiative) is a public-private partnership managed by the African Union Development Agency (AUDA-NEPAD).

The initiative works to restore deforested and degraded lands by planting trees, encouraging natural regeneration, and restoring mangroves, wetlands, and grasslands.

The objective is to strengthen the resilience of landscapes, increase biodiversity, improve food and water security, create jobs, and build stronger and more sustainable economies.

In December 2021, at the end of the first six years of AFR100 implementation, local country commitments exceeded the initial target of 100 million hectares.

The list below details domestic commitments for the 32 African countries involved.

⊕ **148**

AFRICAN COUNTRY	REFORESTATION COMMITMENT	AFRICAN COUNTRY	REFORESTATION COMMITMENT
Benin	0.50	Mozambique	1
Burkina Faso	5.00	Namibia	0.07
Burundi	2.00	Niger	3.2
Cameroon	12.00	Nigeria	4
Chad	1.4	Republic of Congo	2
Central African Republic	3.5	Rwanda	2
Côte d'Ivoire	5.00	Senegal	2
Democratic Republic of Congo	8.00	Sierra Leone	0.7
Ethiopia	15.00	South Africa	3.6
Ghana	2.00	Sudan	14.6
Guinea	2.00	Swaziland	0.5
Kenya	5.1	Tanzania	5.2
Liberia	1.00	Togo	1.4
Madagascar	4.00	Uganda	2.5
Malawi	4.5	Zambia	2
Mali	10.00	Zimbabwe	2

Total commitment of 127,770 millions of hectares of land

When the last tree is cut, the last fish is caught, and the last river is polluted; when to breathe the air is sickening, you will realize, too late, that wealth is not in bank accounts and that you can't eat money.

— Alanis Obomsawim

Leaders Championing Climate Policy

Apolitical.co is an education platform founded by mission-driven entrepreneurs and used by 100,000+ government employees. Prior to 2020, the site compiled a list of 100 leaders influencing climate change. In the extract below, we have listed Apolitical's top 20 personalities and 24 political leaders influencing climate policy and climate change. (The other 56 were artists, youth activists, and NGO and international personalties covered by other articles in the Almanac.)

⊕ **147**

TOP 20 PERSONALITIES

Alexandria Ocasio-Cortez United States	As the youngest US congresswoman in 2019, she was also the lead advocate for the Green New Deal, a proposed economic stimulus program to address climate change, which could lead the US infrastructure transition into renewable energy.
Anne Hidalgo France	As the mayor of Paris, Hidalgo hosted the Climate Summit for Local Leaders in December of 2015. She was also elected to the chairship of C40 cities, an initiative that unites 90 leading global cities on climate change.
Anthony Nyong	He leads the African Development Bank Group's efforts to transition Africa to low-carbon and climate resilience. He coordinated the New Deal on Energy for Africa, an initiative to allow energy access to all Africans by 2025.
Bill McKibben United States	His 1989 book "The End of Nature" is considered the first mainstream book on climate change. He was the winner of the Gandhi Prize and the co-founder of 350.org, an international campaign operating in 188 countries in opposition to new coal, oil, and gas projects.
Catherine McKenna Canada	As Minister of Environment and Climate Change, she worked on the Paris Agreement and secured Canada's first plan in conjunction with provinces, territories, and Indigenous peoples to address climate change and develop a clean economy.
David Attenborough England	His documentary "Blue Planet II" accelerated plastic recycling. He was influential in campaigns like World Wildlife Fund's push to protect Borneo's rainforest. He addressed the UN Climate Change Summit in 2018.
Fatih Birol Turkey	As Executive Director of the International Energy Agency, he led a comprehensive modernization program, reaching out to countries like India and Brazil. He serves on the UN Secretary-General's Advisory Board on Sustainable Energy for All.
Greta Thunberg Sweden	She is known for her strikes for climate outside the Swedish parliament, demanding governments reduce carbon emissions in accordance with the Paris Agreement.

Hilda Heine Marshall Islands	Heine was the chair of the Climate Vulnerable Forum, an alliance of 48 countries under climate change threat. During her tenure as president of the Marshall Islands, she committed to becoming carbon-neutral by 2050.
Hoesung Lee Korea	An endowed chair professor at the Korea University Graduate School of Energy and Environment, his research includes the economics of climate change, energy, and sustainable development.
Jennifer Morgan Netherlands	She is a former executive director of Greenpeace International, a leading environmental NGO that campaigns on issues like climate change, deforestation, and nuclear power. Her current role is Special Envoy for International Climate Action in the German Foreign Ministry.
Josefa Leonel Correia Sacko Angola	She is an African Agronomist and commissioner for rural economy and agriculture of the African Union Commission. She is renowned for her speechees at the African Development Bank and World Trade Organization amongst others.
Katherine Hayhoe United States	She is an atmospheric scientist and co-authored the book, "A Climate for Change: Global Warming Facts for Faith-Based Decisions." She also co-authored the 3rd National Climate Assessment of 2014.
Marina Silva Brazil	From humble beginnings in a rubber-tapping community, she became the Minister of Environment and reduced deforestation by nearly 60 percent while establishing the Amazon Fund.
Michael Bloomberg United States	As mayor of New York City, his methods cut the city's carbon footprint by 19 percent. He highlighted the role of cities in fighting climate change as the chair of the C40 Climate Leadership Group from 2010-2013. He co-authored "Climate of Hope: How Cities, Businesses, and Citizens Can Save the Planet."
Michael Mann United States	A professor of atmospheric science at Pennsylvania State University and a director of the Penn State Earth System Science Center. He authored several books on climate change and cofounded the science website realclimate.org.
Patricia Espinosa Mexico	She was the executive secretary of the UN Framework Convention on Climate Change.
Pope Francis Vatican City	In 2015 he issued the first papal encyclical on climate change, care for the environment, and sustainability.
Saleemul Hug Bangladesh	He is a scientist, a founding director of the International Center for Climate Change and Development, and a senior fellow at the International Institute for Environment and Development. He was the lead author for the Intergovernmental Panel on Climate Change.
Xie Zhenhua China	As the special representative on climate change for China, he coordinated an agreement between China and the US on reducing carbon emissions and gathered political support for the adoption of the Paris Agreement. As the Minister of Environmental Protection, he advocated for clean air, resource conservation, and sustainable development.

Continued

TOP 24 POLITICAL LEADERS

Al Gore
United States
Former Vice President

A former Vice President, his work in 1992 resulted in the introduction of the UNFCCC. In 2005, he founded the Climate Reality Project, which unites global activists. In 2007, he received the Nobel Peace Prize for his work on global climate change.

Bernie Sanders
United States

A committed environmentalist, Senator Sanders served on the Environment and Public Works committee. With California Senator Barbara Boxer, he introduced the Climate Protection Act to tax carbon and methane emissions. In 2007, he co-wrote the Green Jobs Act and is a strong proponent of the Green New Deal.

Bryony Worthington
United Kingdom

A member of the House of Lords, she played a key role in the Friends of the Earth "Big Ask" campaign, calling for new climate change law. She was the lead author of the UK's Climate Change Act of 2008, introducing "carbon budgets."

Carlos Manuel Rodriguez
Costa Rica

When he served as the Environment and Energy Minister from 2002-2006, he pioneered incentivizing farmers and landlords to use their land for environmental good and helped preserve ecosystems. He also played a key role in curbing unsustainable fishing practices globally.

Caroline Lucas
England

Lucas is an internationally recognized speaker on climate change and is considered an influential member of the global green movement. She became the UK's first Green MP.

Debbie Raphael
United States

She is a scientist and public servant. She was Director of the San Francisco Department of the Environment and responsible for the introduction of original policies on rooftop solar panels and electric vehicle chargers and set an ambitious goal of becoming carbon neutral by 2050. She also serves as President of the Urban Sustainability Director's Network.

Elizabeth May
Canada

She became the first Canadian Green Party candidate to be elected to parliament in 2011. She authored eight books and also served on the board of the International Institute for Sustainable Development.

Harsh Vardhan
India

He launched the anti-plastic bag campaign in 2008 in Delhi and the "Green Shopper Campaign" promoting business partnerships with manufacturers of environment-friendly products.

Jay Inslee
United States

Known as a leading thinker on renewable energy, as governor he made Washington a top state for renewable energy and electric vehicles. He co-authored the book "Apollo's Fire: Igniting America's Clean-Energy Economy." He also cofounded the bipartisan US Climate Alliance to uphold US goals as per the Paris Agreement.

Jøgen Abildgaard
Denmark

Abildgaard is the executive climate project director for the Copenhagen 2025 Climate Plan initiative that aims to make Copenhagen the world's first carbon-neutral city by 2025. He was also the former Danish Minister for Environment and Energy.

Katarina Schulze
Germany

As a German politician and co-leader of the Bavarian Green Party, her policies are focused on social sustainability, European integration, and the development of stringent environmental rules.

Li Ganjie
China

As the youngest appointed Minister of Ecological Environment, he transitioned 35 Chinese cities from gas to electricity in 2018.

Mark Carney England	Carney raised awareness of the economic threat of climate change to the finance sector. He also proposed the appointment of a senior executive responsible for climate change threats at banks and insurers to manage climate change. This resulted in banks being held responsible for environmental actions.
Maricio Rodas Ecuador	As the mayor of Quito, he initiated a sustainable metro system project. In 2016, he hosted the UN Conference of Urban Sustainable Development, Habitat III.
Mohamed Sefiani Morocco	As mayor of Chefchaouen, he committed the city to sustainability. He is president of the Moroccan Association of Eco-Cities. He is also a member of the Intermediate Cities Board and the Global Covenant of Mayors for Climate and Energy.
Mohammed Adjoin Sowah Ghana	As mayor of Accra, he steered plans to make Accra the cleanest city in Africa. His policies are about improving the cleanliness and health of the city, including better sanitation, improved waste management, and a "polluter pays" scheme to curb bad environmental behavior.
Mukta Tilak India	As mayor of Pune municipality, her leadership banned all dumping of solid waste and took measures to improve air quality. Her work resulted in Pune receiving the Innovative Policy award at the 2018 Climate and Clean Air Awards held at the Global Climate Action Summit.
Piyush Goyal India	He led the electrification of nearly 18,000 villages. He championed India's renewable energy expansion program (the world's largest). He received the Carnot Prize in 2018 for energy policy contributions.
Rick Kriseman United States	As mayor of St Petersburg in Florida, he led the city in launching an innovative plan to reduce air pollution, implemented a community solar program, and increased renewable energy financing.
Sergio Bergman Argentina	A rabbi and Environment Minister, he advocated an ethical approach to environmental policies. He oversaw the 2017 G20 Sustainability Working Group, tasked with rebuilding the discussion around climate change for developing nations.
Sheldon Whitehouse United States	The junior senator for Rhode Island and member of the Senate Environment and Public Works Committee, he supported initiatives to cut carbon pollution and protect air and water. He founded the Senate Oceans Caucus promoting creative bipartisan policy solutions that protect oceans, coasts, people, and economies.
Sonam Phuntsho Wangdi Bhutan	He helped Bhutan become a leader in mitigating climate change by achieving a net-zero carbon footprint—one of the few countries to achieve that goal.
Teresa Ribera Spain	She proposed the country's first climate plan that requires corporations to report their carbon footprint and aims to achieve carbon neutrality for Spain by 2050.
Tri Rismaharini Indonesia	She transformed the city of Surabaya from one suffering from pollution and congestion into one championing sustainability and rich green spaces. Her leadership created eleven landscaped parks. She was recognized as one of world's 50 greatest leaders by Fortune magazine.

Global Climate Youth Activists

Greta Thunberg is not the only climate activist of her generation. Her public appearances and school strike have focused attention on the climate, and there are many other youth leaders and groups working on this issue as well.

The Fridays For Future (FFF) movement was founded in August 2018. The goal of FFF is to put moral pressure on policymakers to listen to the scientists and take necessary action to limit a global rise in temperatures.

Earth.org, an environmental news & data platform aiming to illustrate the repercussions of climate change, compiled a list of 10 other global youth climate activists who participated in the UN Climate Change Conference of Youth in November 2021.

🌐 **139**

Less than nine months after Greta Thunberg stood with her sign at the Swedish Parliament, more than one million people participated in the School Strike for Climate.

YOUTH ACTIVIST	BIO
Xiuhtezcatl Martinez United States Against use of fossil fuels	An environmental activist, hip-hop artist, and advocate for Indigenous and marginalized communities. Has made speeches at the UN in languages including English, Spanish, and his native Nahuatl.
Nyombi Morris Uganda Against forest logging	A climate justice campaigner who is not easily deterred by physical threats or the suspension of his Twitter account. He protects his country's forest in light of Uganda's vulnerability to extreme climate events.
Licypriya Kangujam India Against air pollution	One of the world's youngest activists, she protested outside the Indian Parliament to make climate change literacy mandatory in schools. She gave six TEDx talks before the age of 10.
Xiye Bastida Mexico Lobbying global climate action by governments	Co-founder of Re-Earth Initiative, organizer of Fridays For Future in NY and committee member of People's Climate Movement, Xiye witnessed the severe impacts of climate change when extreme floods hit her hometown San Pedro Tultepec.
Lesein Mutunkei Kenya Mission: Planting trees	Founder of Trees4Goals, Lesein plants 11 trees for every football goal he scores. He encourages schools and football clubs to be more sustainable and is working on expanding his campaign across Africa.
Luisa Neubauer Germany Lobbying for climate policies to surpass the Paris Agreement goals	Often referred to as "German Greta Thunberg." Campaigned for the University of Gottingen to stop investing in fossil fuel industries and endorsed policies such as degrowth. She is a member of the youth wing of the German Greens Party.
Autumn Peltier Canada Fighting for clean drinking water for First Nation communities	At her UN General Assembly speech in 2019, Peltier famously remarked, "I've said it once, and I'll say it again—we can't eat money or drink oil."
Ella and Amy Meek United Kingdom Fighting against plastic pollution and waste	The two sisters started "Kids Against Plastic" in 2016. They launched campaign initiatives with over 1000 schools and over 50 businesses and festivals. They have also given multiple talks and published *Be Plastic Clever* in 2020.
Kevin J. Patel United States Fighting against air pollution and impacts of climate change in LA	Suffered severe heart issues as a child due to dirty air. Patel is co-deputy partnerships director for Zero Hour, lead organizer for Youth Climate Strike LA, and founder of OneUpAction International.
Qiyun Woo Singapore Raising awareness of complex climate issues and sustainability-related causes	An environmental activist and artist, Woo influences the circular economy, sustainable finance, environmental policies and ecology with her educational artwork. She works with a wide range of stakeholders and discusses economic models and ecofeminism.

NGOs Working to Address Climate Change Around the World

There are now tens of thousands of organizations around the world working on climate change issues. Here is a sample of NGOs, in no particular order. India, with a population of more than a billion, is well represented on this list.

ORGANIZATION	DESCRIPTION
International Institute of Health and Hygiene (IIHH)	Collaborates with national and international funding agencies to develop software and hardware related to health, hygiene, and sanitation.
The Energy and Resource Institute (TERI)	Dedicated to sustainable and inclusive development through conservation of energy and innovative waste management.
VATAVARAN	A confederation of twelve India-based organizations promoting the welfare of animals and people as well as the reduction and recycling of waste.
Vanari	Combats climate change through forest management and sustainable development in rural India.
Uttarkhand Seva Nidhi Paryavaran Shiksha Sansthan (USNPSS)	Presents environmental education programs in schools and villages in the mountain region of Uttarakhand, India.
Orissa Environmental Society	Conducts research and publishes materials related to protection and conservation of natural resources and the environment.
Ladakh Ecological Development Group (LEDeG)	Organizes sustainable development programs in the towns and remote villages of Ladakh, India.
Kalpavriksh	Promotes environmental awareness, campaigns, litigation, and research. Confronts the state through measures ranging from protest letters to street demonstrations.
Green Future Foundation	Studies ecosystems and sustainable livelihood techniques to conserve the landscapes of India.
Shakti Sustainable Energy Foundation	Aids the design and implementation of policies that promote clean power, energy efficiency, and sustainable transport.
Navdanya Trust	Has created more than 150 community seed banks across India.
M S Swaminathan Research Foundation (MSSRF)	Seeks to help Indian farmers and fisherfolk to incorporate modern science and technology.
Indian Council of Forestry Research and Education (ICFRE)	Has patented a number of innovations related to forestry that support rural and tribal livelihoods.

Development Alternatives (DA)	Creates innovations in the areas of building, water management, and renewable energy to reduce poverty and rejuvenate natural ecosystems in developing regions of India.
Environics Trust	Supports communities affected by mining and disasters as well as marginalized and displaced populations in the Himalayan mountains and coastal regions.
C.P.R. Environmental Education Centre (CPREEC)	Educates local communities in South India about the environment with a special focus on schoolteachers, women, and children.
Centre for Science and Environment (CSE)	Conducts lobbying and research related to sustainable development.
Centre For Environmental Studies (CES)	Promotes environmental education, awareness, training, and research.
G. B. Pant Institute of Himalayan Environment and Development	Works toward the conservation of natural resources and the promotion of environmentally sound development in the Indian Himalayan Region.
National Institute of Occupational Health (NIOH)	Aims to improve the management of occupational health risks in India.
Centre for Media Studies (CMS)	Dedicated to equitable development and responsive governance in India.
Indian Environmental Society (IES)	Promotes grassroots, community-based environmental conservation initiatives.
Wildlife Trust of India (WTI)	Projects include preventing wild animal deaths resulting from trains and teaching shark hunters to be conservationists.
World Wide Fund (WWF-India)	Works to ensure conservation of the biodiversity of India.
Wildlife Protection Society of India (WPSI)	Collaborates with state governments to curb poaching and illegal wildlife trade.
Satpuda Foundation	Protects the largest tiger habitat in the world through work at the policy and grassroots levels.
Balajee Sewa Sansthan (BSS India)	Strives for hygiene, clean water, sanitation, and social and cultural equality among the underprivileged.
ASSIST	Provides sustainable solutions for the conservation, usage, and maintenance of water sources.
Haritika	Develops natural resource management solutions and infrastructure to fight the effects of climate change on the rural poor of Bundelkhand, India.
Technology Informatics Design Endeavour (TIDE)	Helps to make rural Indian women economically independent through the installation of low-cost, fuel-efficient stoves and other technology.
Abhinav	Supports the advancement of rural people in Uttar Pradesh, India, especially promoting clean water, water conservation, and use of technology in agriculture.
Greenpeace	Global network that seeks to expose global environmental problems through peaceful protest and creative communication.
Earth Institute Center for Environmental Sustainability	Promotes sustainability through wide-ranging collaborations and fosters understanding of ecology and the importance of biodiversity.
Earth Island Institute	Sponsors projects addressing environmental leadership and other conservation efforts, as well as legal advocacy for environmental issues.

Earth Justice	An environmental law organization representing clients in legal cases related to climate change, renewable energy, wildlife, and human health.
Environmental Defense Fund	Battles urgent threats to the environment.
Fauna and Flora International	Protects against biodiversity loss worldwide through investments, local solutions, and technology.
Naturefriends International	Comprised of about 45 environmental organizations with an emphasis on tourism and cultural heritage.
Global Footprint Network	Gathers data on individual humans' ecological footprints.
International Union for Conservation of Nature	Defends nature by uniting government and societal groups.
The Nature Conservancy	Conserves lands worldwide through direct efforts and partnerships, focusing on biodiversity concerns and climate change.
Natural Resources Defense Council	Unites a large citizen member base with scientists, lawyers, and advocates to protect the environment.
Wetlands International	Conserves wetlands around the world.
World Agroforestry (ICRAF)	Applies knowledge of trees to improve food security and sustainability.
World Wildlife Fund	Works with communities to preserve their local natural resources and aims to adjust policies in favor of sustainability actions.
The Environmental Foundation for Africa	Works to protect and restore the environment in West Africa.
350.org	Aims to end the use of fossil fuels and transition to renewable energy by building a global grassroots movement.
SustainableEnergy	Works to ensure access to affordable, reliable, sustainable, and modern energy for all.
Blue Ventures	Partners with local communities to design and scale marine conservation and fisheries management.
Ukraine Nature Conservation Society	Promotes recycling and environmental education at schools, in local communities, and among government authorities.
Conservation Through Public Health	Enables humans to live safely with gorillas and other wildlife.
The Finnish Association for Nature Conservation	The largest environmental protection and nature conservation organization in Finland.
Emirates Environmental Group	Organizes clean-up drives and waste collection facilities across the Emirates and educates the public on conservation, sustainability, and recycling.
International Centre for Integrated Mountain Development	Shares and implements innovative solutions for sustainable mountain development with the eight countries of the Hindu Kush Himalaya.
Bund für Umwelt und Naturschutz Deutschland	Supports renewable energy and lobbies in Brussels and Berlin regarding environmental and climate policy.
Corporate Europe Observatory	Seeks to research and expose corporate influence in EU policy.
Comunidad Inti Wara Yassi	Dedicated to environmental education and the care of sick, mistreated, and abandoned wildlife.

Clean Air Network	Encourages the public to speak out and support government measures that could improve the quality of air in Hong Kong.
Bellona Foundation	Employs ecologists, scientists, engineers, economists, lawyers, and journalists to identify and implement solutions to environmental problems.
Ancient Forest Alliance	Committed to protecting British Columbia's old-growth forests in areas where they are scarce while ensuring sustainable forestry jobs there.
Haribon Foundation	A nature conservation organization in the Philippines dedicated to sustainability and empowering community participation in environmental stewardship.
Casa Pueblo	A Puerto Rican community organization formed in response to a proposed mining project, it seeks to practice and promote the responsible use of the land's ecosystems and resources.
Pro Natura	The oldest nature preservation organization in Switzerland.
International Energy Agency	Facilitates the global dialogue on energy, providing analysis, data, and policy recommendations that allow countries to move toward secure and sustainable energy.
David Suzuki Foundation	Conducts research, education, and policy analysis to promote conservation and protection of the natural environment in Canada.
The Climate Reality Project	A collective of activists, cultural leaders, organizers, scientists, and storytellers committed to building an inclusive and sustainable future.
C40	97 global cities addressing climate change by developing and implementing policies and programs that generate measurable reductions in both greenhouse gas emissions and climate risks.
Friends of the Earth International	An international network of organizations focused on environmentalism and human rights.
Rainforest Alliance	An international organization working at the intersection of business, agriculture, and forests to build an alliance to protect forests and improve the livelihoods of farmers and communities.
Green Cross	Responds to the combined challenges of security, poverty, and environmental degradation through dialogue, mediation, and cooperation.
World Resources Institute	Global research organization that works with governments, businesses, multilateral institutions, and civil society groups to develop practical solutions that improve people's lives and protect nature.
Citizen's Climate Lobby	A grassroots, nonpartisan climate change advocacy organization in the United States.
Climate Alliance	One of the largest European city networks dedicated to climate action.
The Carbon Underground	Mitigates climate change through soil regeneration and restorative agriculture.
Earthworks	Safeguards land from the consequences of new developments for mining and energy production.

⊕ **135**

Leaders of Civics Programs Addressing Environmental Issues

These change makers have been recognized by the North American Association for Environmental Education and are Fellows in a leadership program addressing issues related to Civics and Environmental Education. The complete list of Fellows working on projects directed towards adults, as well as youth, can be viewed online.

⊕ **143**

LEADER	INITIATIVE	OBJECTIVE	TARGET AUDIENCE
Mandy Baily	Community Voices, Informed Choices	Training extension agents to host "facilitated, value-based, and inclusive discussion" in communities.	Extension agents
Ramona Big Eagle	Tower to Table for Food Security, Education, and Sustainability	Create an intergenerational experience centered around food, nutrition, gardening, and entrepreneurship.	Seniors and children in underserved communities; individuals living in food deserts
Cesar Almeida	Dancing for Environmental Justice	Connect BIPOC artists and educators to green spaces in Chicago through performing arts events.	BIPOC artist community, natural sites, nature centers, parks, heritage sites, botanical gardens
Siya Aggrey	Integrating Environmental Education within Community-based Disease Surveillance System in Mountain Communities of Elgon Region (ECSEMER)	Enhance resilience in mountain communities of Mount Elgon, whose agricultural livelihood is threatened by recurring disruptive events influenced by climate change.	Communities of Mount Elgon, local high schools, and health providers
Shannon Francis	Mycelium Healing Project	Filter pollutants in the soil, air, and water in Commerce City, CO.	Latinx communities in Commerce City
Shougat Nazbin Khan		Develop solar photovoltaic-powered vending carts for street vendors.	Street vendors
Matt Kirchman	Benchmarks for Environmental Literacy in Museums	Develop and publish resources for museum professionals that help them incorporate environmental literacy into their exhibition practices.	Museum professionals

Judith Morales	Plastic Pollution Awareness Program	Establish awareness and support behavior change around plastic consumption.	University students
Kevin O'Connor	Keepers of Our Place: Community Environmental Monitoring Project	Work with neighboring schools and communities to address environmental issues in their region through place- and land-based education.	Community members responding to social, geographical, environmental, and economic issues
Melanie Schikore	Neighbor2Neighbor	Create "permaculture precincts" to introduce neighbors to each other; establish communication and experiences that help communities move towards sustainable behavior.	Residents
Olivia Walton	Sustainable Food for Freedom City	Create interpretive signage for a local fish market encouraging sustainable fishing practices, environmental stewardship, and creative ideas about how the market space can be used as an inclusive community space.	The community of Frederiksted, US Virgin Islands
Lisa Yeager	Climate Conversations: Improvising Our Way to Improved Civic Engagement	Create a climate conversation toolkit and train-the-trainer resources for volunteers in informal learning environments.	Volunteers in informal learning environments

Top Global Universities for Studying the Environment, Ecology, and Climate

RANK	INSTITUTION	
1	Wageningen University and Research Center Data	Netherlands
2	Stanford University	United States
3	Harvard University	United States
4	University of California, Berkeley	United States
5	Swiss Federal Institute of Technology Zurich	Switzerland

The field of environment and ecology includes subjects such as environmental health, environmental monitoring and management, and climate change. These are the top global institutions of higher learning focused on the study of the environment and ecology.

As published by *US News and World Report*, this list of the top global institutions of higher learning focused on the study of the environment and ecology was based on data from the Web of Science for the five-year period from 2015-2019.

⊕ **144**

Change happens by listening and then starting a dialogue with the people who are doing something you don't believe is right.
— Jane Goodall

Influential Artists and Climate

For generations, artists in all media have commented and influenced discussions in our culture. The following is a list of artists selected by Christie's, Artsy, Huffington Post and others for their influential work about the environment, climate change, conservation, and sustainability.

🌐 **142**

ARTIST	SHORT BIO
Agnes Denes (Hungary, USA), conceptual and land art	Agnes transforms abandoned spaces into natural oases. Her "Wheatfield - A Confrontation" was cultivated on a landfill site opposite the World Trade Center in Manhattan. The crop produced 450kg of grain that was shipped to 28 cities as a secondary exhibit. Also influential are "Rice/Tree/Burial" in New York and "Tree Mountain - A Living Time Capsule" in Finland.
Aida Sulova (Kyrgyzstan), street art	Pasting photographs of large open mouths onto city trash cans, she depicts the world's garbage making its way into humans.
Allison Janae Hamilton (USA), sculpture, installation, photography, video	Her immersive work shows how natural and climate-related disasters expose social and racial inequities. For example, her work "The peo-ple cried mer-cy in the storm" pays tribute to black migrant workers killed during hurricanes in the 1920s.
Amanda Schachter and **Alexander Levi** (Spain, USA), performance architecture	Co-founders of SLO Architecture, this husband-and-wife team created Harvest Dome 2.0, which shows that garbage can be reused in beautiful ways.
Andreas Gursky (Germany), photographer	Influential works include "Oceans" (rising sea levels) and "Bangkok" (waterway pollution).
Andy Goldsworthy (England), land art, sculpture	Long on the cutting edge of land art, his installations use material from the surrounding landscape and only exist until erased by nature. Sculptures have been built in Yorkshire Sculpture Park and near the Storm King Art Center, New York.
Barry Underwood (USA), multimedia	He creates light installations within a landscape to draw attention to issues such as light pollution and deforestation.
Cai Guo-Qiang (China), sculpture, conceptual art	His work shows the paradigm shift of nature now existing at the mercy of humans. Influential works include "Ninth Wave" (a fishing boat filled with sculptures of endangered species), "The Bund Without Us," and "Silent Ink."
Chris Jordan (USA), photography	Chris addresses issues such as consumption and waste with images depicting massive dumps of cell phones, circuit boards, etc.

Christo (deceased) and **Jeanne-Claude** (Bulgaria, France, USA), environmental sculpture	They used massive land installations in public spaces to draw attention to the natural world. For their "Surrounded Islands," 40 tons of garbage was cleared in the process of construction.
Daan Roosegaarde (Netherlands), installation	An innovative designer, his prominent works include "Smog" and "Seeing Stars," which address issues of air and light pollution.
David Buckland (UK), film	David is the founder of Cape Farewell, a non-profit in which artists, scientists and activists take on cultural and ecological projects towards a more sustainable future.
David Maisel (USA), photography	His large-scale images depict aerial views of landscapes transformed by water reclamation, logging, military tests, and mining.
Denilson Baniwa (Brazil), paint, photography, performance	As an urban artist, he raises awareness of environmental and Indigenous issues in the Amazon, such as pesticide contamination and toxic mining residue on Indigenous lands. He was a participant in the Arctic Amazon Symposium 2019 on common climate strategies.
Edward Burtynsky (Canada), photography	Winner of a TED prize in 2005, the recipient of Sony's World Photography Award 2022, and contributor to the Anthropocene Project, his large-scale landscape images show the devastation of human activity on the Earth's surface.
El Anatsui (Nigeria), sculpture	Since the 1970s, he has worked with scrap and found material to highlight issues of colonialism, extraction, waste, and renewal.
Gabriel Orozco (Mexico), multimedia, sculpture	His exhibits showcase found items and trash in the landscape. Notably, "Sandstars" draws attention to industrial and commercial pollution in Isla Arena, Mexico.
John Akomfrah (England, Ghana), film, videography	A member of the Black Audio Film Collective, he describes his influential video installation "Purple" as "a person of color's response to the Anthropocene."
John Sabraw (USA), paint	An activist and environmentalist, John is pursuing a fully sustainable practice. He makes paints from iron oxide in the runoff water from abandoned coal mines. Current projects include "Anthrotopographies" and "Hydrophilic."
Justin Brice Guariglia (USA), conceptual art, fine art	Drawing attention to climate issues such as melting ice sheets through collaboration with NASA, the Climate Museum, the New York City Mayor's Office, and others, his solar-powered "Climate Signals" raises awareness of rising sea levels.
Leah Anthony (Canada), zines	From the Nak'azdli Band, she is a winner of the Fraser Valley Indigenous Youth Climate Art Contest for her artwork zine UNEVEN GROUND: laying down complex root systems.
Lisa K. Blatt (USA), photo, video, installation	Lisa's art, made in extreme landscapes like Antarctica, plays with the edges of perception, making the invisible impacts of climate change visible. Influential works include "Clearest Lake in the World" and a collection of "Heatscapes," shown during the 2018 UN Global Climate Action Summit.
Luzinterruptus (Spain), installation	An anonymous collective making urban art in public spaces, their prominent works include "Labyrinth of Plastic Waste" and "Plastic Islands."

Mary Mattingly (USA), photography, sculpture, installation, performance	Mary's work questions systemic issues that impact the environment, such as access to clean water. "Swale" is an ongoing, interactive public art installation attempting to change policy and reconnect communities to local food sources.
Mathilde Roussel (France), sculpture	A series of living grass and recycled material sculptures in human form, "Lives of Grass" draws attention to food cycles, abundance, and scarcity.
Mel Chin (USA), conceptual art	Significant works include "Revival Field," which uses plants to extract heavy metal from soil; "Operation Paydirt;" and "Unmoored," which uses augmented reality to imagine Times Square under water. Mel is a member of Cape Farewell.
Naziha Mestaoui (Belgium), architecture	She is best known for "One Beat One Tree," a digital interactive display installed at COP21 that supports reforestation throughout the world.
Noŋgirrŋa Marawili (Australia), bark paint and printmaking	A senior Madarrpa artist, she uses ink from discarded cartridges on natural materials to document culture, history, and the environment.
Olafur Eliasson (Denmark, Iceland), multimedia and large scale installations	Olafur is a UN Goodwill Ambassador for Climate Action and co-founder of solar energy company Little Sun, which is working to displace fossil fuel energy in communities without electricity. Prominent works include "Ice Watch," which was displayed at the Paris Climate Conference, and "Earth Perspectives," which celebrates the fiftieth anniversary of Earth Day.

Photo detail courtesy of Lisa K. Blatt

Paulo Grangeon (France), sculpture	He is best known for his travelling exhibit "Pandas on Tour" in collaboration with the WWF, in which 16,000 papier-mâché pandas shed light on animal endangerment.
Rachel Sussman (USA), photography	She has spent ten years photographing the oldest organisms on Earth, some up to 80,000 years old. The photographs have been published in a book, "The Oldest Living Things in the World."
Random International (Germany), experimental	Their digital exhibit "Rain Room" invites audiences to experience control over rain and participate in a future-stabilized environment.
Shepard Fairey (USA), street art	An environmental activist since the mid-1990s, his extensive work is shown worldwide from MoMa in New York to the V&A London. His "Earth Crisis" was installed at the Eiffel Tower for COP21 to symbolize harmony and climate threats.
Tomas Sarceno (Argentina), architect	Tomas is a member of Aerocene, a collaboration that strives for "an ethical collaboration with the atmosphere and the environment, free from borders, free from fossil fuels." His work "Museo Aero Solar" is a floating, solar-powered museum made of recycled plastic.
Xiuhtezcatl Martinez (USA), hip-hop	An activist since childhood, he has been a climate spokesperson at the UN General Assembly and UN Summit Brazil and is the youth director of Earth Guardians.

Principles for Responsible Investing

Kofi Annan, former United Nations Secretary-General, invited a group of the world's largest institutional investors to develop the Principles for Responsible Investing (PRI). The principles were launched in April 2006 at the New York Stock Exchange.

PRI's mission is to create an economically efficient and sustainable global financial system that rewards long-term responsible investment and benefits the environment and society as a whole.

This goal will be achieved by encouraging adoption and implementation of the following principles:

Principle 1: Incorporate ESG (Environmental, Social & Governmental) issues into investment analysis and decision-making processes.

Principle 2: Be active owners and incorporate ESG issues into ownership policies and practices.

Principle 3: Seek appropriate disclosure on ESG issues by the entities that receive investment.

Principle 4: Promote acceptance and implementation of the Principles within the investment industry.

Principle 5: Work together to enhance effectiveness in implementing the Principles.

Principle 6: Report activities and progress towards implementing the Principles.

Since the PRI was launched, it has amassed over 4,600 signatories. They are categorized as investment managers, asset owners, or service providers.

The PRI showcases signatories at the cutting edge of responsible investment in their Leaders Group, based on reporting response and assessment data.

⊕ **140**

I believe hypocrisy is unavoidable. You simply can't live in this world without sometimes crossing lines, like taking a plane. It's difficult to live a pure life in an impure situation. Try to avoid hypocrisy, but it's not the worst sin.

Compromise is unavoidable, and in fact, should be encouraged. There's a lot of purism and hair-shirt wearing in the environmental movement that we have to forego. If we can't work with everybody and anybody, then we have failed.

— Brian Eno

UN PRI 2020 leaders group

SIGNATORY	SIGNATORY CLASS	SIZE US$ BN	COUNTRY
ACTIAM	Investment Manager	50 - 249.99	Netherlands
AkademikerPension	Asset Owner	10 - 49.99	Denmark
Allianz SE	Asset Owner	≥ 250	Germany
AMP Capital Investors	Investment Manager	50 - 249.99	Australia
AP2	Asset Owner	10 - 49.99	Sweden
APG Asset Management	Investment Manager	≥ 250	Netherlands
Australian Ethical Investment Ltd.	Investment Manager	1 - 9.99	Australia
Aware Super	Asset Owner	50 - 249.99	Australia
AXA Investment Managers	Investment Manager	≥ 250	France
Bridges Fund Management	Investment Manager	0 - 0.99	UK
Brunel Pension Partnership (BPP)	Asset Owner	10 - 49.99	UK
Candriam Investors Group	Investment Manager	50 - 249.99	Luxembourg
CBUS Superannuation Fund	Asset Owner	10 - 49.99	Australia
CDC - Caisse des dépôts et Consignations	Asset Owner	50 - 249.99	France
Charter Hall Group	Investment Manager	10 - 49.99	Australia
Church Commissioners for England	Asset Owner	10 - 49.99	UK
Dexus Investment Manager	Investment Manager	10 - 49.99	Australia
Environment Agency Pension Fund	Asset Owner		UK
ESG Portfolio Management Investment Manager	Investment Manager	0 - 0.99	Germany
Ilmarinen Mutual Pension Insurance Company	Asset Owner	50 - 249.99	Finland
Legal & General Investment Management	Investment Manager	≥ 250	UK
Lendlease	Investment Manager	10 - 49.99	Australia
Manulife Investment Management	Investment Manager	≥ 250	Canada
Mirova	Investment Manager	10 - 49.99	France
Natixis Assurances	Asset Owner	50 - 249.99	France
Neuberger Berman Group LLC	Investment Manager	≥ 250	United States
New Zealand Superannuation Fund	Asset Owner	10 - 49.99	New Zealand
Nuveen, a TIAA Company	Investment Manager	≥ 250	United States
Payden & Rygel	Investment Manager	50 - 249.99	United States
Robeco	Investment Manager	50 - 249.99	Netherlands
State Street Global Advisors (SSGA)	Investment Manager	≥ 250	United States
Stichting Pensioenfonds ABP	Asset Owner	≥ 250	Netherlands
Swedfund International AB	Asset Owner	0 - 0.99	Sweden
The International Business of Federated Hermes	Investment Manager	10 - 49.99	UK
Universities Superannuation Scheme - USS	Asset Owner	50 - 249.99	UK
Varma Mutual Pension Insurance Company	Asset Owner	50 - 249.99	Finland

A Corporate Race for Sustainability

The World Benchmarking Alliance (WBA) is a not-for-profit organization founded in 2018 that measures and ranks the world's 2000 most influential companies on their contribution towards achieving the UN's Sustainable Development Goals (SDGs). WBA's goal is to inspire a corporate race to the top.

One of the seven benchmarks produced is the Climate and Energy benchmark, ranking 450 of the world's most influential companies in high-emitting sectors against the Paris Agreement and SDG 13.

It uses a holistic approach to assess a company's readiness to transition to a low-carbon economy, applying quantitative and qualitative evaluations of a company's climate strategy, business model, investments, operations and greenhouse gas emissions management.

The tables here list the top 10 automotive, electric utilities, and oil & gas sector companies based on the 2021 benchmarks.

🌐 141

TOP 10 COMPANIES IN THE AUTOMOTIVE SECTOR	HEADQUARTERS	SCORE (OUT OF 100)
Tesla	United States of America	71
Renault	France	62
Volkswagen	Germany	52
BYD	China	50
BMW	Germany	49
Daimler	Germany	48
General Motors Corporation	United States of America	48
SAIC Motor	China	46
Guangzhou Automobile Group	China	45
Tata Motors	India	44

TOP 10 COMPANIES IN THE ELECTRIC UTILITIES SECTOR	HEADQUARTERS	SCORE (OUT OF 100)
Ørsted	Denmark	96
SSE	United Kingdom	84
E.ON	Germany	79
Vattenfall	Sweden	78
Energias de Portugal	Portugal	77
Enel	Italy	74
Iberdrola	Spain	70
Électricité de France	France	67
Engie	France	67
Xcel Energy	United States of America	64

TOP 10 COMPANIES IN THE OIL AND GAS SECTOR	HEADQUARTERS	SCORE (OUT OF 100)
Neste	Finland	57
Engie	France	57
Naturgy Energy	Spain	45
Eni	Italy	44
bp	United Kingdom	43
Total	France	41
Repsol	Spain	38
Equinor	Norway	38
Galp Energia	Portugal	36
Royal Dutch Shell	Netherlands	34

*Since changing its focus to preserving the environment,
the clothing company Patagonia has consistently grown and increased profits.*

They offer recycled goods, free repairs, and a focus on re-use instead of consumption.

Patagonia's Corporate Manifesto

Build the best product

Our criteria for the best product rests on function, repairability, and, foremost, durability. Among the most direct ways we can limit ecological impacts is with goods that last for generations or can be recycled so the materials in them remain in use. Making the best product matters for saving the planet.

Cause no unnecessary harm

We know that our business activity—from lighting stores to dyeing shirts—is part of the problem. We work steadily to change our business practices and share what we've learned. But we recognize that this is not enough. We seek not only to do less harm, but more good.

Use business to protect nature

The challenges we face as a society require leadership. Once we identify a problem, we act. We embrace risk and act to protect and restore the stability, integrity, and beauty of the web of life.

Not bound by convention

Our success—and much of the fun—lies in developing new ways to do things.

DON'T BUY
THIS JACKET

> "There are no simple solutions. The constant increase in temperature, climate change, the progressive reduction of water resources, phenomena which were once hard to see but now evident and accelerating, challenge us and require us legislators to make quick, consistent and ambitious choices.
>
> I think it is clear to everyone that the time of denial, the time of delays, the time of resistance to changing a lifestyle that is proving unsustainable for the planet, are behind us.
>
> The scale of today's challenge is unprecedented. The COVID-19 pandemic is showing us that in cases like these borders no longer matter because no one is safe from these phenomena. We are therefore called to make a joint and shared effort that has few precedents in the history of humanity to completely rethink our lifestyle, to find and implement new and more efficient environmental technologies.
>
> We should have taught our children and grandchildren that the gift that was given to us, our world, does not have infinite resources and is far more fragile than we thought. Too many times the opposite has happened. Credit must be given to the new generation of young people, who have grasped the urgency for a change of course earlier and better than us."

— David Sassoli

Ranking the Greenest Companies

According to the 2021 *Corporate Knights* Global 100, these are the 10 most sustainable corporations in the world. Seven of these companies have committed to achieving net zero and/or working to cap the increase in global temperature at 1.5°C/2.7°F.

Factors in each company's overall score include:

1. **Energy Productivity:** Energy used minus renewable energy and/or certified renewable energy credits generated.
2. **Greenhouse Gas Productivity:** Emissions caused by company-controlled and/or company-owned sources plus emissions resulting from electricity, steam, heat, and/or cooling services purchased by the company.
3. **Water Productivity:** Water used or withdrawn and not returned to its source for reuse.
4. **Waste Productivity:** Waste created minus waste recycled.

5. **Pollutant Productivity:** Output of volatile organic compounds, nitrogen oxides, sulfur oxides, and particulate matter emissions.
6. **Supplier Sustainability Score:** Largest publicly-listed supplier (measured by total expenditure) scored using the Corporate Knights Global 100 formulas minus the Supplier Sustainability Score.
7. **Sustainability Pay Link:** Formal monetary incentives for senior executives hitting sustainability goals.
8. **Sanctions Deductions:** Deductions for companies whose ratio of paid fines/penalties/settlements to total revenue exceeded those of their industry group peers from 2016-2019.
9. **Clean Revenue:** Proceeds yielded by goods and services with a positive environmental impact.
10. **Clean Investment:** Corporate outlays for goods and services with a positive environmental impact.

⊕ **136**

2022 RANK	COMPANY	COUNTRY	CLIMATE COMMITMENTS	SCORE
1	Vestas Wind Systems A/S	Denmark	1.5°C, SBTi	A+
2	Chr Hansen Holding A/S	Denmark	1.5°C, SBTi	A
3	Autodesk Inc	United States	SBTi	A
4	Schneider Electric SE	France	1.5°C, SBTi	A
5	City Developments Ltd	Singapore	1.5°C, SBTi	A
6	American Water Works Company Inc	United States		A
7	Orsted A/S	Denmark	1.5°C, SBTi	A-
8	Atlantica Sustainable Infrastructure PLC	United Kingdom	SBTi	A-
9	Dassault Systemes SE	France	1.5°C, SBTi	A-
10	Brambles Ltd	Australia	1.5°C, SBTi	A-

1.5°C (Business Ambition for 1.5°C): a global coalition (set up by UN Global Compact, Science Based Targets initiative, and We Mean Business) enlisting companies to commit to capping the global temperature increase at 1.5°C/2.7°F.

SBTi (Science Based Targets initiative): Companies committed to SBTi are reducing emissions at scale to help halve greenhouse gas emissions by 2030 and reach net zero by 2050.

Resources

Get engaged

The Educators Guide

The Educators Guide to *The Carbon Almanac* will help educators confidently use the Almanac to help students address climate-related subjects. This free resource is filled with suggestions that can easily be incorporated into lessons, discussions, and activities.

⊕ **177**

THE EDUCATORS GUIDE INCLUDES

*A quick-start guide to the almanac, frameworks for leading discussions about climate science, activities that will help you use The Carbon Almanac, and direct links to additional sources. Find it at **thecarbonalmanac.org/177***

"Whenever I want my mom to play fort, I just turn on news about climate change."

YOU CAN MAKE A DIFFERENCE

Visit **www.thecarbonalmanac.org** and sign up for **The Daily Difference**, a free email that will connect you with our community. Every day, you will join thousands of other people connecting around specific actions and issues that will add up to a significant impact.

Read, Watch, Listen, Act

Climate change is causing economic, social, and cultural upheavals, and they're all going to compound. Creators around the world are highlighting the stories and issues around these shifts. Here are some of the projects that are worth exploring. Find out more at **www.thecarbonalmanac.org/resources**.

BOOKS / NONFICTION

Intersectional Environmentalist, Leah Thomas. 2022.

Basics on the ways privilege and systemic racism impact environmental issues and activism in minority and underrepresented communities, as well as tips on leading inclusively. Written by the founder of the BIPOC-centered climate justice site intersectionalenvironmentalist.com. For teens and up.

Green Ideas series, various authors. 2021.

*Twenty short books by environmental leaders such as Greta Thunberg (**No One Is Too Small To Make a Difference**), Michael Pollan (**Food Rules**), and Rachel Carson (**Man's War Against Nature**). For teens and up.*

The Future Earth, Eric Holthaus. 2020.

A hopeful view of what the world would look like if we achieved net zero. For those who want a dose of optimism.

How to Avoid a Climate Disaster, Bill Gates. 2021.

The business magnate's survey of current emissions-reducing technologies and what innovations are still needed, along with his accessible plan to hold communities, corporations, and governments accountable for making critical changes. For teens and up.

The New Climate War, Michael E. Mann. 2021.

Battle strategies for cutting through climate denial and pressuring corporations and governments to end fossil fuel use. For anyone wanting to make a bigger impact.

The Physics of Climate Change, Lawrence M. Krauss. 2021.

The science of global warming presented in an accessible way. For readers wanting the basics.

Regeneration: Ending the Climate Crisis in One Generation, Paul Hawken et al. 2021.

The Drawdown author's fresh takes on saving a world running out of time. For those seeking inclusive initiatives.

Saving Us, Katharine Hayhoe. 2021.

A climate scientist's tips for persuasive environmental discussions with people on all sides of the issues. For adults looking to improve their advocacy.

Speed & Scale, John Doerr. 2021.

Hard-nosed venture capitalist tactics for reaching net zero by 2050. For fans of business-based action plans.

Value(s): Building a Better World for All, Mark Carney. 2021.

The former banker's solutions to climate change (and other systemic global issues), rooted in maximizing benefits for the many and not the few. For readers versed in economics and environmental policy.

All We Can Save: Truth, Courage, and Solutions for the Climate Crisis, edited by Ayana Elizabeth Johnson and Katherine Wilkinson. 2020.

*Hopeful essays and poetry written by women leading the green movement and curated by marine biologist Johnson and **Drawdown** (see below) contributor Wilkinson. The companion site, allwecansave.earth, offers resources for reader groups and for managing emotions around climate change. For teens and up.*

The Future We Choose, Christiana Figueres and Tom Rivett-Carnac. 2020.

A cautionary but positive book about climate change and the fate of humanity written by two of the UN's lead negotiators during the Paris Agreement. For adults.

The Story of More, Hope Jahren. 2020.

An impassioned plea to understand climate change and take action. For adults.

The Circular Economy: A User's Guide, Walter R. Stahel. 2019.

An approachable look at securing sustainable development in different sectors and communities. For business and political leaders.

The End of Ice: Bearing Witness and Finding Meaning in the Path of Climate Disruption, Dahr Jamail. 2019.

Environmental stakes brought into high relief against the backdrop of the author's travels through Alaska's Denali Mountains, the Amazon rainforest, and Australia's Great Barrier Reef. For adults.

There is No Planet B, Mike Berners-Lee. 2019.

A generalist book about how to avoid climate disaster. For adults.

Climate: A New Story, Charles Eisenstein. 2018.

An argument for seeing trees, oceans, and other elements of the natural world not as potential carbon storage, but as sacred and meaningful forces in and of themselves. For adults.

Read, Watch, Listen, Act *continued*

Farming While Black, Leah Penniman. 2018.

Both a food-growing guide and a manifesto to ending racism in the agricultural industry. The companion website farmingwhileblack.org links to Soul Fire Farm, the James Beard Award-winner's Afro-Indigenous community farm. For adults.

Ground Truth: A Guide to Tracking Climate Change at Home, Mark L. Hineline. 2018.

Tips on paying attention to the shifts in nature occurring around us. For adults.

What We Know about Climate Change,
Kerry Emanuel. 2018.

MIT Press's update to its 85-page guide (first printed in 2007) to the basic science behind this pressing issue. For adults.

Drawdown: The Most Comprehensive Plan Ever Proposed to Reverse Global Warming, edited by Paul Hawken. 2017.

Heavily-researched catalog of carbon interventions that companies, communities, families, and governments must conduct to combat climate change. Its companion website offers additional options. For those who want to take action.

Energy and Civilization, Vaclav Smil. 2017.

A thorough history of society, its energy sources, and their consequences. For readers who enjoy somewhat technical writing.

The Great Derangement, Amitav Ghosh. 2016.

An unpacking of the conflicting and baffling complexities of the fossil fuel economy by a Booker Prize nominee. For adults.

Who Really Feeds the World?, Vandana Shiva. 2016.

The award-winning scientist and activist's argument that solutions for agricultural sustainability will come from local small-scale farming practices. For adults.

Learning to Die in the Anthropocene: Reflections on the End of a Civilization, Roy Scranton. 2015.

Unsparing reflections on our present and future if we do nothing, written by an Iraq War veteran. For adults. Includes graphic war analogies.

The Mushroom at the End of the World: On the Possibility of Life in Capitalist Ruins, Anna Lowenhaupt Tsing. 2015.

A book on sustainability and what life grows in the wake of industrial activity as told through the matsutake mushroom, said to be the first thing to grow after the Hiroshima bombings. For adults.

Braiding Sweetgrass: Indigenous Wisdom, Scientific Knowledge, and the Teaching of Plants, Robin Wall Kimmerer. 2013.

A modern touchstone reminding us to live according to the truth that we are part of—not separate from—nature. For teens and up.

The Sixth Extinction, Elizabeth Kolbert. 2014.

The author's vision of a sixth mass extinction that has been set into inexorable motion. For adults.

To Cook a Continent: Destructive Extraction and the Climate Crisis in Africa, Nnimmo Bassey. 2012.

The Nigerian architect and activist's analysis of how plundering of Africa for fossil fuels has accelerated global warming's effects there. For adults.

The Gort Cloud, Richard Seireeni. 2009.

Strategies for leveraging brands by tapping into the green community's matrix of NGOs, advocacy groups, social networks, business alliances, and the like. For marketers and corporate leaders.

BOOKS / PHILOSOPHY / INSPIRATIONAL

Blind Spots, Max H. Bazerman and Ann E. Tenbrunsel. 2011.

A discussion of ethical failures and flawed decision-making. Although not specifically about climate change, it outlines ways to plan and execute solutions more effectively. For activists, policymakers, and corporate executives who want to turn the tide.

How to Blow Up a Pipeline: Learning to Fight in a World on Fire, Andreas Malm. 2021.

Not an eco-terrorist handbook, but rather a Swedish ecology professor's clarion call to confront the major fossil fuel players. For adults.

Zen and the Art of Saving the Planet,
Thich Nhat Hanh. 2021.

Stirring meditations and appeals for action from the Zen master and climate activist. For teens and up.

Small is Beautiful: Economics As If People Mattered,
E.F. Schumacher. 2011.

The economist's refutation of "bigger is better," especially with regard to fossil fuels. Written at the peak of the 1970s energy crisis. For adults.

This Is Not a Drill, Extinction Rebellion. 2019.

Simple tips for civil disobedience tactics such as blocking roads, occupying bridges, and feeding protesters. Readers can get involved at rebellion.global. For young adults and up.

Desert Solitaire, Edward Abbey. 2011.

A posthumous reprint of the park ranger's passionate musings on nature and humanity's disregard for its destruction of the planet. First released in 1968. For adults.

Eat Like a Fish, Bren Smith. 2019.

A case for how ocean-farming seaweed and shellfish will feed the world and clean its waters. For those seeking responsible food choices or alternatives to fishing and traditional land farming.

BOOKS / BIOGRAPHY / MEMOIR

Finding the Mother Tree, Suzanne Simard. 2021.

A paean to the wisdom and social interconnectedness of forests. For fans of science-based memoir.

Warmth: Coming of Age at the End of Our World, Daniel Sherrell. 2021.

Hope, despair, and perseverance from the frontlines of the climate movement. For adults.

Horizon, Barry Lopez. 2019.

The National Book Award-winner's musings on climate change woven through his explorations of Kenya's deserts, Antarctica, the Galapagos, and more. For adults.

The Wizard and the Prophet, Charles C. Mann. 2018.

A science lesson disguised as a history lesson about two scientists whose work laid the foundations for opposing schools of thought — "Cut back!" vs. "Innovate and grow!" — about the environment. For adults.

The World-Ending Fire, Wendell Berry. 2017.

Odes to rural living and the urgent need for sustainability from the longtime Kentucky farmer and essayist. For teens and up.

Beyond the Horizon, Colin Angus. 2010.

The author's credo of zero-emissions travel as exemplified by his journey to become the first person to complete a self-powered circumnavigation of Earth. For adventure lovers.

No Impact Man: The Adventures of a Guilty Liberal Who Attempts to Save the Planet, Colin Beavan. 2009.

The record of a yearlong experiment for the author, his wife, and young daughter as they try to make zero impact on the environment while living in Manhattan. For adults.

BOOKS / FICTION

Bewilderment, Richard Powers. 2021.

The story of an astrobiologist widower and his neurodivergent, nine-year-old son trying to navigate an environmentally endangered world. Shortlisted for the 2021 Booker Prize. For adults.

How Beautiful We Were, Imbolo Mbue. 2021.

An elegiac novel about a fictional African village and its people's battle to wrest their land away from an American oil company responsible for killing their children and ravaging their soil. For adults.

Once There Were Wolves, Charlotte McConaghy. 2021.

A tale about one woman's desperate fight to protect the Scottish Highlands and her beloved wolves from their human enemies. For teens and up.

The Ministry for the Future, Kim Stanley Robinson. 2020.

An optimistic yet unsettling vision about how humans can build a different sort of future before it's too late. Essential reading.

The Waste Tide, Chen Qiufan. 2019.

A dystopian story set in the near future about a Chinese "waste girl" who takes a low-wage job manually recycling toxic e-waste and ends up leading her fellow trash workers in a bloody revolution; translated from the 2013 Chinese edition. For adults.

The Overstory, Richard Powers. 2018.

Powerful examples of environmental activism and resistance found in the journeys of its nine main characters. Winner of the 2019 Pulitzer Prize for Fiction. For adults.

The Ministry of Utmost Happiness, Arundhati Roy. 2017.

A seething epic tackling everything from India's impending vulture extinction to its deforestation, tainted rivers, and growing slums from a Booker Prize-winning author. For adults.

The Collapse of Western Civilization, Naomi Oreskes and Erik M. Conway. 2014.

A fictional but science-based account of Earth in 2393 and how centuries of drought, ice melt, and willful inaction rendered the planet unrecognizable. For adults.

The Windup Girl, Paolo Bacigalupi. 2009.

Far-reaching science fiction about climate, Thailand, and human perseverance. For adults.

Salvage the Bones, Jesmyn Ward. 2011

The two-time National Book Award-winner's story of a working-class Black family in Mississippi during the short span of days before and immediately after Hurricane Katrina. Conveys the devastation extreme weather has on marginalized populations. For adults.

Ishmael Series trilogy (Ishmael, The Story of B, and My Ishmael), Daniel Quinn. 1992-1997.

Three novels employing magical realism about humanity's role as stewards of nature. Features Ishmael, a telepathic gorilla who teaches two pupils how to save the world. Not written for kids, but suitable for preteens and up.

Read, Watch, Listen, Act *continued*

BOOKS / POETRY

The Glass Constellation, Arthur Sze. 2021.

New and reprinted poems wrestling with life under the looming shadow of climate change from a National Book Award winner and Pulitzer Prize finalist. For teens and up.

Ultimatum Orangutan, Khairani Barokka. 2021.

A defiant collection scrutinizing environmental injustice and its colonialist roots. For young adults and up.

Habitat Threshold, Craig Santos Perez. 2020.

Stanzas that range from awestruck to mournful to instructive as the Guam native ponders the destructiveness of global industry and the ecological fate of his homeland. For teens and up.

BOOKS / CHILDREN'S

Dr. Wangari Maathai Plants a Forest, Rebel Girls. 2020.

A biography about the Kenyan environmentalist who became the first African woman to win a Nobel Peace Prize; part of the popular edutainment series. For ages 5-13.

We Are Water Protectors, Carole Lindstrom and Michaela Goade. 2020.

The Caldecott Award-winning tale of an Ojibwe girl who leads resistance efforts against an oil pipeline. For ages 3-6.

The Magic School Bus and the Climate Challenge, Joanna Cole. 2010.

Simple global warming concepts for kids from the hit edutainment series. For ages 7 and up.

Our Changing Climate, UNICEF Zimbabwe. 2017.

A free online book introducing the issues at hand, illustrated with examples from daily life in Zimbabwe; can be found at unicef.org/zimbabwe. For ages 11-12.

Understanding Photosynthesis with Max Axiom, Super Scientist, Liam O'Donnell. 2007.

A graphic novel about how plants use carbon to make food. For ages 8-14.

The Lorax, Dr. Seuss. 1971.

Dr. Seuss's personal favorite, which encourages kids to take individual responsibility for protecting the environment. Also a 2012 movie directed by Chris Renaud. For all ages.

LECTURES

Climate Justice Can't Happen without Racial Justice, David Lammy. 2020.

The Member of Parliament for Tottenham, England, on the importance of inclusion and BIPOC climate leadership. For preteens and up.

Community Investment Is the Missing Piece of Climate Action, Dawn Lippert. 2021.

Tips on getting citizen enrollment for climate action. Part of the TED Talks Daily series. For teens and up.

The Standing Rock Resistance and Our Fight for Indigenous Rights, Tara Houska. 2018.

Houska is an Ojibwe attorney and environmental and Indigenous rights advocate. This is a firsthand account of the standoff against the Dakota Access oil pipeline, as well as North America's rampant Indigenous erasure that has allowed fossil fuel companies to exploit tribal lands. For preteens and up.

The Quest for Environmental and Racial Justice for All: Why Equity Matters, Dr. Robert Bullard. 2017.

The urban planning and environmental policy professor's signature lecture at MIT about the causes of and solutions to the racial segregation of America's pollution. For preteens and up.

Breaking the Tragedy of the Horizon, Mark Carney. 2015.

Climate risk explained in the language of economics. For adults.

A 40-Year Plan for Energy, Amory Lovins. 2012.

The scientist and renewable energy advocate's free-market proposal to wean the US off oil and coal by 2050 at a $5 trillion savings—without the need for new federal law.

Global Warming, Global Threat, Dr. Michael McElroy. 2003.

Audiobook lecture series by a Harvard professor on the science of the greenhouse effect, failures to address the rise in emissions, and who's responsible for taking crucial next steps. For adults.

PODCASTS

Bioneers: Revolution from the Heart of Nature, Neil Harvey.

Sustainability stories about climate justice, food, farming, Indigenous knowledge, restricting corporate power, and youth activism told with depth and empathy. For all ages.

Black History Year: "Environmental Racism: A Hidden Threat with Dr. Dorceta Taylor," Jay Walker.

A dialogue with the environmental studies professor about the interconnectedness of racism, economic injustice, and the impact of climate change, as well as the steps for BIPOC communities to take charge of their own fates. For teens and up.

The Carbon Copy, Stephen Lacey.

Weekly news analyses with experts, journalists, business leaders, and other guests about current events and their climate impacts. For teens and up.

Catalyst, Shayle Kann.

Interviews with experts about decarbonization and climate technology solutions. For tech enthusiasts.

Climate One, Greg Dalton.

In-depth talks with activists, influencers, and decision-makers in front of live audiences. For teens and up.

Drilled, Amy Westervelt.

A true-crime style podcast with seasons on corporate-financed climate denial and local communities seeking justice from fossil fuel companies.

How to Save a Planet, Alex Blumberg.

Self-described climate nerds unafraid to get silly (voice impressions of recycling bins, anyone?) to get listeners energized about climate activism. For all ages.

Outrage and Optimism, Christiana Figueres, Tom Rivett-Carnac, and Paul Dickinson.

Free-ranging conversations with leading climate change figures. For teens and up.

Planet Money: "Waste Land," Sarah Gonzalez and Laura Sullivan.

An episode about the lie of plastic recycling perpetuated by manufacturers and oil companies so they could continue doing business as usual. For teens and up.

Political Climate, Brandon Hurlbut, Shane Skelton, and Julia Pyper.

A bipartisan podcast on energy and environmental politics in America. For teens and up.

The Response, Tom Llewellyn.

Deep dives into how different communities recover and establish resiliency after natural disasters. For teens and up.

Scene On Radio: Season 5, The Repair, John Biewen and Amy Westervelt.

The two-time Peabody-nominated podcast on the colonizing Western forces that caused climate change. Insight from places such as Jakarta, Nigeria, and Bangladesh that have been devastated by fossil fuel damage. For teens and up.

Sourcing Matters, Aaron Niederhelman.

Discussions on where our food comes from, how sourcing impacts climate change, and what reforms are possible. For teens and up.

Sustainababble, Oliver Hayes and David Powell.

British improv comedy meets environmental research. For teens and up.

Sustainability Defined, Jay Siegel and Scott Breen.

Humorous, listener-friendly analysis of different aspects of the environmental movement. For teens and up.

Think: Sustainability, Marlene Even and Sophie Ellis.

Practical suggestions for adopting greener consumer habits through the lens of inclusivity. For teens and up.

The Yikes Podcast, Mikaela Loach and Jo Becker.

A UK-based intersectionality-focused program to hearten those anxious about issues of climate change and social justice. For teens and up.

MOVIES

Don't Look Up, Adam McKay. 2021.

A satirical, star-studded climate change allegory about a comet on the verge of wiping out the Earth and the scientists desperate to convince the media and the government the threat is real. Rated R.

Vanishing Lines, Fancy Tree Films. 2021.

An 18-minute documentary about a planned European ski resort expansion that would destroy major glaciers. For preteens and up.

The Year Earth Changed, David Attenborough. 2021.

Remarkable footage of the return of clearer skies, greener lands, and healthier wildlife after COVID-19 necessitated worldwide shutdowns and isolation in 2020. Rated PG.

David Attenborough, A Life on Our Planet, Alastair Fothergill, Jonathan Hughes, and Keith Scholey. 2020.

The longtime natural historian's firsthand account of humanity's monumental impact on the wild. Rated PG.

Kiss the Ground, Joshua Tickell and Rebecca Harrell Tickell. 2020.

Scientists and activists on reversing climate change by returning carbon dioxide and microorganisms to the soil. For preteens and up.

Plastic Wars, Rick Young. 2020.

The investigation by NPR and the PBS show "Frontline" into how the plastic industry used recycling as a marketing ploy to boost plastic demand and sales, making the trash problem much worse. Rated PG.

Read, Watch, Listen, Act *continued*

Our Planet, various directors. 2019.

A Netflix and World Wildlife Fund series celebrating the stunning flora, fauna, and landscapes of Earth. Narrated by David Attenborough. Rated G.

An Inconvenient Sequel: Truth to Power, Bonni Cohen and Jon Shenk. 2017.

A follow-up to An Inconvenient Truth tracking Al Gore as he advocates for renewable energy investment and the completion of the Paris Agreement. Rated PG.

Beyond Climate, Ian Mauro. 2016.

An award-winning documentary centered on British Columbia's environmental efforts to counteract wildfires, glacier erosion, flooding, and oil pipelines. For preteens and up.

To the Ends of the Earth, David Lavallee. 2016.

Stories of conservationists and environmental leaders standing up for lands being destroyed by fossil fuel extraction in Alberta, Utah, and the Arctic. Narrated by Emma Thompson. For preteens and up.

Nowhere to Run: Nigeria's Climate and Environmental Crisis, Dan McCain. 2015.

A film about Nigeria's dire levels of drought, desertification, and violent land conflicts brought on by deforestation and fossil fuel consumption. Hosted by the late Nigerian environmental activist Ken Saro-Wiwa Jr. For teens and up.

Who Killed the Electric Car?, Chris Paine. 2006.

The surprising and fraught history of the electric car. Narrated by Martin Sheen. Rated PG.

Interstellar, Christopher Nolan. 2014.

A sci-fi film about humans trying to escape to another planet as Earth is plagued by sandstorms and global crop disease in the year 2067. Rated PG-13.

Mission Blue, Robert Nixon and Fisher Stevens. 2014.

An Emmy-winner for outstanding editing in documentary or long form. Chronicle of marine biologist Sylvia Earle's quest to create national park-like "hope spots" in the ocean to preserve biodiversity and counteract climate damage. For teens and up.

Aluna: An Ecological Warning by the Kogi People, Alan Ereira. 2012.

The remote Colombian mountain tribe's entreaties to protect the environment. A sequel to the 1990 documentary, From the Heart of The World: Elder Brother's Warning. For teens and up.

Chasing Ice, Jeff Orlowski. 2012.

Environmental photographer James Balog's time-lapse footage of global warming's destruction of massive ancient glaciers. Rated PG-13.

Food, Inc., Robert Kenner. 2009.

An exposé of the heavy health and environmental tolls exacted by the cost-cutting assembly-line practices of the global food industry. Its companion book, Food, Inc.: A Participant's Guide, unpacks how human diets affect climate change. Rated PG.

Home, Yann Arthus-Bertrand. 2009.

Aerial views of Earth's wonders as well as the damage humans have caused to the natural world. Narrated by Glenn Close. For preteens and up.

WALL-E, Andrew Stanton. 2008.

An Academy Award-winning animated film about how love and hope spark one trash-compacting robot's quest to resurrect Earth from the 29th-century wasteland it has become. Rated G.

An Inconvenient Truth, Davis Guggenheim. 2006.

The Oscar-winning documentary about Al Gore's crusade to inform people about global warming. Includes the slideshow presentation he used during that campaign. Rated PG.

Manufactured Landscapes, Jennifer Baichwal. 2006.

A film about photographer Edward Burtynsky's trip to China to create compelling images of large-scale industrial structures and their environmental impacts on local terrain. For preteens and up.

FernGully: The Last Rainforest, Bill Kroyer. 1992.

An animated musical about magic fairies fighting industrial destruction of their rainforest home. Teaches kids about the importance of caring for Earth. Rated G.

WEBSITES

(visit thecarbonalmanac.org/resources for the links)

The Arctic Cycle

Live performances and storytelling intended to spark climate conversations and spur people to action. For adults.

Artists & Climate Change

A blog encouraging artists to create and write about global warming-themed work to inspire connections with the green movement. An initiative of The Arctic Cycle. For anyone looking for creative expressions of climate issues.

Artists for Climate: The Climate Collection

A selection of open-licensed, environment-themed digital illustrations conveying optimism and action. For everyone, especially educators, graphic designers, and students.

Cambridge Institute for Sustainability Leadership

An organization partnering with corporations, governments, and financial institutions to build a thriving global green economy driven by the UN's Sustainable Development Goals. For businesses, policymakers, and people looking for climate action training.

Canary Media

One of the leading news companies focused on the transition to a decarbonized economy and society. Funded by Rocky Mountain Institute. For adults.

Climate Reality Project

Al Gore's international organization for training climate leaders. For teens and adults wanting a structured approach to participating in the movement.

The Conversation: Environment + Energy

Environmental articles written by academics and edited by journalists; US and international editions available. For teens and up.

David Suzuki Foundation

A conservation group partnering with businesses and governments to fix key environmental problems through scientific research, education, and policy analysis. For adults.

Earthjustice

A free environmental law service— the ACLU for climate change. For anyone unable to afford environmental legal help and experts willing to volunteer.

Earthwatch

A global nonprofit connecting volunteers with scientists to do environmental research that will help safeguard the planet. For the science-minded and corporate/educational partners.

Ellen MacArthur Foundation

A charity whose mission is replacing the "take, make, waste" mentality with one focused on ending pollution, circulating goods, and replenishing nature. For consumers, businesses, and policymakers.

Environmental Voter Project

A nonpartisan nonprofit which identifies non-voting environmentalists and turns them into voters. For young adults and up.

First Nations Climate Initiative

A British Columbia-based forum founded by the Lax Kw'alaams, Metlakatla, Nisga'a, and Haisla First Nations to combat climate change and decarbonize the economy while working to end poverty and create environmental leaders in Indigenous communities. For anyone interested in learning more about Indigenous-led climate action.

Fridays for Future

A site for the international youth movement of school strikes to compel adults to take responsibility for climate change. Founded by Greta Thunberg. For students.

The Great Green Wall

An African project to plant a nearly 5,000-mile-long wall of trees through the continent as a way to mitigate climate change and drought while providing jobs and food security for locals. For all ages.

Green 2.0

Watchdogs calling out inequities in the environmental movement. Publishes an annual diversity report card on green NGOs and foundations. For anyone interested in inclusivity.

Inside Climate News

Pulitzer Prize-winning, nonpartisan environmental journalism. For preteens and up.

Juma Institute

An organization founded by Indigenous Brazilian activist Juma Xipaia to safeguard the Amazon rainforest and those working to preserve it. Some posts are in Portuguese. For teens and up.

Post Carbon Institute

A research group offering data and analysis on energy conservation, sustainability, and ecological resilience. For adults.

Reasons To Be Cheerful: Climate + Environment

A good-news site on innovative solutions to pressing climate issues being implemented by small communities, cities, and national governments. Stories come from diverse places such as the Congo, Samsø Island, and the village of Kamikatsu, Japan. For teens and up.

The Rocky Mountain Institute

A nonpartisan nonprofit of interdisciplinary experts working with legislators, corporations, and institutions to decarbonize energy systems worldwide. For those in business, finance, and energy sectors.

Sierra Club

A grassroots organization guarding everyone's right to a healthy planet. Founded in 1892. For all ages.

350.org

An international group working to end all fossil fuel use. Founded by award-winning environmentalist Bill McKibben. For students, activists, and anyone seeking news about climate change action.

Read, Watch, Listen, Act *continued*

Women's Earth Alliance

An empowerment community for women worldwide offering technical and tactical training for leading green initiatives as well as a supportive network of donors, peers, and mentors. For women.

Work On Climate

An activist Slack channel for making connections, exchanging knowledge, creating companies, and finding paid or volunteer environmental work. For teens and adults who want to get involved.

World Benchmarking Alliance

A group working to incentivize top corporations to meet the UN's Sustainable Development Goals by measuring their SDG contributions. For academic and research institutions, business platforms, financial institutions, governmental entities, NGOs, and sustainability consulting firms.

World Wildlife Fund

A leading conservation NGO protecting Earth's natural resources for animals and communities. For everyone.

ONLINE RESOURCES

Breathe This Air

Experts on the toxic effects of Louisiana's plastic factories on nearby Black communities. Features environmental professor Dr. Beverly Wright, eco-justice organizer Dante Swinton, and Goldman Environmental Prize winners Sharon Lavigne and Prigi Arisandi. For teens and up.

Can You Fix Climate Change?

A clear and entertaining summary of the many complex layers involved in solving the emissions problem from the popular Kurzgesagt channel (the German equivalent for "in a nutshell"). For preteens and up.

Causes and Effects of Climate Change

National Geographic's primer on the roots and ramifications of increasing emissions for the environment and humans. For all ages.

Climate Victory Gardens

A video on gardening to restore carbon to the soil and counteract atmospheric CO_2. Features Rosario Dawson and LA-based guerrilla gardener Ron Finley. For all ages.

Ecological Footprint Calculator

A colorfully illustrated personal quiz that computes how many Earths it would take to power the world if everyone lived like you. For all ages (although kids may need help answering the questions).

Just Have a Think

Weekly sustainability solutions researched and presented by concerned citizen Dave Borlace (based in the UK). For teens and up.

Kimiko Hirata, 2021 Goldman Environmental Prize, Japan

An overview of the Japanese activist's accomplishments which include the cancellation of 13 planned coal plants in her homeland. For teens and up.

Studio B: Unscripted—Kumi Naidoo and Winona LaDuke

A two-part conversation between the longtime environmental activists. Naidoo is the global ambassador for Africans Rising for Justice and a former Greenpeace executive director, and LaDuke is an Ojibwe farmer and economist. For teens and up.

The Tipping Point: Climate Change

The BBC's concise presentation on global warming. For teens and up.

Wangari Maathai and the Green Belt Movement

The Nobel Peace Prize-winning activist on her Nairobi-based grassroots NGO, which empowers rural communities economically by teaching them sustainable, regenerative agricultural methods. Since its 1977 inception, GBM has been responsible for planting millions of trees and training thousands of women as foresters, beekeepers, and food processors. For teens and up.

NEWSLETTERS

Heated, Emily Atkin.

Billed as "accountability journalism for the climate crisis." For readers seeking fiery takes from a criticially-acclaimed climate journalist.

Minimum Viable Planet, Sarah Lazarovic.

A hope-filled weekly newsletter about fighting climate change. For anyone interested in taking action.

Getting Started with Climate Action

Piotr Drozd, the COO of Leaders for Climate Action, posted a list on LinkedIn that expanded into a crowdsourced effort as comments were contributed. Find the links related to each of these at ⊕ **162**.

INDIVIDUAL / CITIZEN

Bark.today *Dutch organization providing research and information to help individuals reduce their ecological footprint and build a biodiverse society*

Count Us In *A mission-based project to inspire 1 billion citizens to significantly reduce their carbon pollution and challenge leaders to deliver bold, global change*

TheClimateSavers *Facilitating partnership and collaboration amongst people committed to decelerating climate change*

CROWDSOURCING SUSTAINABILITY

Do Nation *A global community making collective pledges to build healthier habits for people and the planet*

Ecologi *An environmental organization providing a subscription model to fund climate action, grow forests, and track actions to reduce global emissions*

Good Empire app *Bringing together collective action around the world aligned with all 17 of the UN's Sustainable Development Goals*

Giki Zero *"Get Informed, Know Your Impact" by calculating and understanding your carbon footprint and learning about meaningful action to take*

Joro app *Empowering consumers to take actions that make a difference and build a climate positive lifestyle*

Klima app *Calculate, offset, and reduce your carbon footprint*

Project Drawdown *Solutions-oriented organization to educate and drive climate action*

UGO *Platform in partnership with Karma Volunteering to connect students with sustainable development projects and initiatives*

UN ActNow *Campaign and app to take action to track lifestyle habits and make changes around 10 key areas*

We Don't Have Time *A review and social media platform to spread climate knowledge and influence businesses, organizations, and public leaders to act on climate change*

Crowdsourcing Sustainability *Community and movement driving sustainable action to reverse climate change*

BUSINESSES

B Corp Climate Collective *B Corps who have committed to net zero by 2030*

Business Declares *Network of businesses declaring a climate and ecological emergency, taking purposeful action to reach carbon neutrality*

Leaders for Climate Action *Community of European entrepreneurs and business leaders accelerating progress towards the Paris Agreement goals*

Planet Mark *Providing certification and solutions for businesses to reach and validate net-zero targets*

Pledge To Net Zero *Environmental industry's global commitment, requiring science-based targets from its signatories to reduce greenhouse gas emissions*

SME Climate Hub *Global initiative to enable SMEs to take climate action and work towards halving emissions by 2030 and net zero by 2050 or sooner*

Tech Zero *Group for tech companies committed to climate action*

The Science Based Targets initiative (SBTi) *Drives ambitious climate action in the private sector by enabling companies to set science-based emissions reduction targets*

The Chambers Climate Coalition *Global forum that offers members actionable, real-world solutions and recommendations on cost-effective, sustainable business practices aligned with Paris targets*

The Climate Pledge *Cross-sector of businesses committed to reaching net-zero carbon by 2040*

B1G1 *Social enterprise helping businesses achieve more social impact by embedding giving activities into everyday business operations*

Compare Your Footprint *Provides a comprehensive carbon footprint calculator and benchmarking tool*

Small99 *Practical guidance for small business owners on getting to net zero*

Sustaineers *Community of business professionals committed to achieving global sustainability goals*

TheGreenShot *App to drive sustainable film production*

Getting Started with Climate Action _continued_

Pawprint _Employee engagement tool which harnesses the energy employees already have to fight climate change and channels it towards their organization's climate targets_

ClimateScape _Open directory of companies, investors, NGOs, and other organizations that support climate solutions_

ACTIVIST / CAMPAIGNER

350 _Global grassroots movement to end reliance on fossil fuels_

Climate Action Network _Global network of NGOs working to promote government and individual action to limit human-induced climate change to ecologically sustainable levels_

The Climate Reality Project _Provides climate education and action-oriented training to create Climate Reality Leaders_

Earth Day Network _Mobilizing civil society to create an environmental movement worldwide_

European Climate Pact _EU initiative inviting people, communities, and organizations to participate in climate action_

Extinction Rebellion _Movement using non-violent direct action and civil disobedience to persuade governments to act justly on the Climate and Ecological Emergency_

Fridays For Future _International movement of students who skip Friday classes to participate in demonstrations to demand action from political leaders_

Rainforest Action Network _Mobilizing collective action to stop deforestation, defund fossil fuels, and support indigenous communities_

SumOfUs _Global community committed to curbing the growing power of corporations_

Sunrise Movement _Youth movement to stop climate change and create millions of good jobs in the process._

KlimaDAO _Digital currency with the goal to accelerate the price appreciation of carbon assets making low-carbon technologies and carbon-removal projects more profitable_

Citizens' Climate Lobby _International grassroots environmental group that trains and supports volunteers to build relationships with their elected representatives in order to influence climate policy_

ENTREPENEUR / INNOVATOR

Carbon13 _Works with founders to build startups that can reduce carbon dioxide equivalent emissions by millions of tons_

Cleantech Open _Largest clean technology accelerator program in the world_

Conservation X Labs _Technology and innovation company that creates solutions to stop the extinction crisis_

Elemental Excelerator _Global nonprofit at the intersection of climate, innovation, and equity_

Katapult _Investment company, focused on highly scalable impact tech startups_

On Deck Build for Climate _Eight-week sprint for experts and operators who want to build a Minimum Viable Product (MVP) in climate tech_

Postcode Lotteries Green Challenge _Sustainability competition for start-ups from Germany, Great Britain, the Netherlands, Norway, and Sweden_

Third Derivative _Open, collaborative climate tech ecosystem that accelerates startups_

Urban Us _Seed stage for startups that upgrade cities for climate change_

Build a Climate Startup _Climate tech venture studio and investor based in Europe, on a mission to eliminate at least one gigaton of carbon dioxide equivalent from annual emissions_

Greentown Labs _Climate tech startup incubator with locations in Boston and Houston_

VertueLab _Nonprofit fighting climate change by providing funding and holistic entrepreneurial support to cleantech startups_

Active Impact Investments _Driving environmental sustainability through profitable investment and accelerating the growth of early-stage climate tech companies_

EMPLOYEE / CONTRACTOR

80,000 Hours _Provides research-based advice on careers that have the largest positive social impact_

Work on Climate _Action-oriented Slack community for people serious about climate work_

Climate People *A sustainable climate recruiting firm*

Climatebase *Leading platform for climate careers*

Conservation Job Board

Green Jobs Network

Escape the City *Purposeful jobs, courses, events, and resources*

Women in Cleantech and Sustainability *Supporting women in pursuit of cleantech and sustainable careers and lifestyles*

Planetgroups *Supporting climate action in the workplace*

Low Carbon Business School *Cohort-based free course for employees (particularly of consumer goods companies) to learn and take climate action in their organizations*

Terra.do *Cohort-based education platform for professionals with the mission of getting 100 million people working on climate change solutions by 2030*

Climate Change AI *An organization composed of volunteers from academia and industry to catalyze impactful work at the intersection of climate change and machine learning*

CITY / REGION / STATE

CityInSight *Enables cities to explore energy and emissions scenarios for policy, finance, and infrastructure*

ClimateView *Swedish climate action technology company that helps cities transform climate planning into progress*

Futureproofed *Helping cities and companies transition to a fossil-free future*

ICLEI ClearPath *An online software platform for completing greenhouse gas inventories, forecasts, climate action plans, and monitoring at community-wide or government-operations scales*

Kausal *Helps cities turn their climate goals into action through a digital platform that enables smarter collaboration around key data*

Resilient Cities Network *Brings together global knowledge, practice, partnerships, and funding to empower members to build safe and equitable cities for all*

Whether you are a business executive, a small farmer, a factory manager, or a concerned individual—the following lists have resources to find more information, take action, and connect with others who share your interest.

Facilities Seeking to Reduce Greenhouse Gas Emissions

These resources provide information for facilities emitting greenhouse gases. They are useful to industries and local residents seeking to better understand climate change. Five are highlighted, with a more complete list at www.thecarbonalmanac.org/resources.

RESOURCE	DESCRIPTION
Science & Innovation	US Department of Energy (US DOE) (energy.gov/science-innovation)	Provides current agency policy and research into renewable energy and carbon capture as undertaken by 17 national labs. Details regarding DOE loans for qualifying projects and available funding opportunities for public/private research.
Energy, Climate Change, Environment	European Commission (ec.europa.eu)	Homepage for the European Commission on Energy, Climate Change, and the Environment. Defines agency labeling and reporting requirements, policies, and targets. Also defines standards with practical advice and insight into project implementation and tools used throughout the region.
Carbon Capture, Utilisation and Storage	International Energy Agency (IEA) (iea.org)	The IEA collects, assesses, and disseminates worldwide energy statistics while offering training and sharing best practices to governments across the globe.
California Air Resources Board (CARB) (arb.ca.gov)	California Air Resources Board (CARB) has overseen one of "the most extensive air monitoring networks" in the world for over 50 years. CARB focuses primarily on pollution from "moving sources" (such as boats, cars, and trucks) while local air quality management districts focus on pollution from "stationary sources."
Greenhouse Gas Reporting Program (GHGRP)	US Environmental Protection Agency (US EPA) (www.epa.gov/ghgreporting)	The GHGRP allows businesses and others to track and compare facilities' greenhouse gas emissions, identify opportunities to cut pollution, minimize wasted energy, and save money. States, cities, and other communities can use the data to find high-emitting facilities in their area, compare emissions between similar facilities, and develop common-sense climate policies.

Local Government Climate Resources

This section includes links to local government organizations that are fighting climate change. Resources include information for policy-makers, conferences, and education materials. Three are highlighted, with a complete list at www.thecarbonalmanac.org/resources.

RESOURCE	DESCRIPTION
Local Governments for Sustainability	ICLEI (iclei.org)	A global network of more than 2,500 local and regional governments committed to sustainable urban development. Active in 125+ countries, ICLEI influences sustainability policy and drives local action for low emission, nature-based, equitable, resilient, and circular development.
Global Covenant of Mayors for Climate & Energy (globalcovenantofmayors.org)	10,000+ cities partner with national and international institutions to tackle climate change through local initiatives, innovative financing models, and sustainable infrastructure.
Rocky Mountain Institute (rmi.org)	RMI is a US-based nonprofit working globally and in partnership to encourage rapid, market-based change in critical geographies.

Transportation Industry & Sustainability

This section includes several governmental as well as industry-specific websites, books, and other resources targeted toward improving sustainability and reducing the carbon impact of transportation companies. Five are highlighted to get you started, with a more complete list at www.thecarbonalmanac.org/resources.

RESOURCE	DESCRIPTION
Research & Technical Resources: Sustainability | American Public Transportation Association (apta.com)	North American public transportation industry group's overview and recommendations for best practices related to sustainability for public vehicles.
State & Local Sustainable Transportation Resources | US Department of Energy (US DOE) (energy.gov/eere/slsc)	A compilation of information and links to programs focused on sustainability in state and local transportation.
Why Freight Matters to Supply Chain Sustainability | US Environmental Protection Agency (US EPA) (epa.gov/smartway)	This article explains how companies involved in the production, distribution, and transportation of goods can help mitigate emissions trends and make a difference.
The Centre for Sustainable Road Freight (csrf.ac.uk)	A collaborative venture between Cambridge University, Heriot Watt University, Westminster University and industry organizations in the freight and logistics sectors, taking a cross-disciplinary approach to reducing the carbon imprint in logistics. Their areas of focus include data collection & analysis, logistics, vehicle systems, energy systems, and strategy.
Greenhouse Gas Reporting Program (GHGRP) | US Environmental Protection Agency (US EPA) (www.epa.gov/ghgreporting)	The GHGRP allows businesses and others to track and compare facilities' greenhouse gas emissions, identify opportunities to cut pollution, minimize wasted energy, and save money. States, cities, and other communities can use the data to find high-emitting facilities in their area, compare emissions between similar facilities, and develop common-sense climate policies.

Business & Investor Sustainability

These resources include links to investor resource groups, transparency guidelines, and corporate reporting standards related to climate change. Resources include information for board members, investors, and activists as well as education materials. Four are highlighted to get you started, with a more complete list at www.thecarbonalmanac.org/resources.

RESOURCE	DESCRIPTION
Asia Investor Group on Climate Change (AIGCC) (aigcc.net)	Launched as part of the Global Investor Coalition. This initiative creates awareness among Asia's asset owners and financial institutions about the risks and opportunities associated with climate change and low-carbon investing.
CDP (formerly Carbon Disclosure Project) (cdp.net)	Nonprofit charity that runs the global disclosure system for investors, companies, cities, states, and regions to measure, disclose, manage, and share environmental information.
Financial Stability Board (FSB) (Fsb.org)	The FSB created the Task Force on Climate-related Financial Disclosures (TCFD) and a set of voluntary disclosure recommendations for use by companies in providing decision-useful information to investors, lenders, and insurance underwriters about the climate-related financial risks that companies face.
Global Reporting Initiative (GRI) (globalreporting.org)	More stakeholder than investor-focused, GRI helps businesses and governments worldwide understand and communicate their impact on sustainability issues such as climate change, human rights, governance, and social well-being. The GRI Sustainability Reporting Standards are developed with multi-stakeholder contributions and rooted in the public interest.

Resources for Climate Entrepreneurs

This section contains a range of resources for individuals interested in starting or working at start-ups dedicated to fighting climate change. It is organized into three sections:

- information (podcasts, blogs, etc.) about climate entrepreneurship
- databases of investors who fund climate entrepreneurs
- examples of companies that are (or began as) start-ups, which are making an impact in various climate-related areas

These companies innovate in a wide variety of areas such as food and textiles, energy, finance, and transportation. Three are highlighted to get you started, with a more complete list at www.thecarbonalmanac.org/resources.

RESOURCE	DESCRIPTION
The Exponential View (exponentialview.co)	A newsletter, podcast, and climate-related jobs board produced by technologist Azeem Azhar. Azhar describes EV as a "transdisciplinary guide to our near future." He explores AI, blockchain, synthetic biology, renewables, and other rapidly evolving fields, and prominently features a weekly "countdown clock" of the approach to 450ppm.
ClimateTechVC (climatetechvc.org)	Run by a diverse group of climate-related investors and entrepreneurs, ClimateTechVC provides an extensive database of venture capital funds, corporate investors, and accelerators focused on climate-related start-ups. There are also newsletters, research insights, and a job board.
Modern Meadow (modernmeadow.com)	Modern Meadow is a private, early-stage biofabrication company that has developed biotechnologies for the creation of leather and other textiles without reliance on animals or fossil fuels. With over 2.3 billion livestock slaughtered every year for their skins, Modern Meadow's alternative approach makes a significant contribution to reducing the global effects of ranching.

Construction Sustainability

These resources provide websites, articles, and tools by construction industry professionals and those who contract with them. The ideas found in these resources apply anywhere, subject to local building codes and standards. Four are highlighted to get you started, with a more complete list at www.thecarbonalmanac.org/resources.

RESOURCE	DESCRIPTION
The World Economic Forum (WEF) (weforum.org)	Information and resources on both climate change and the construction industry.
World Business Council For Sustainable Development (WBCSD) (wbcsd.org)	A global consortium led by CEOs across industries to promote sustainable development.
International Energy Agency (IEA) (iea.org)	The IEA collects, assesses, and disseminates worldwide energy statistics, offers training, and shares best practices with governments across the globe.
National Pollutant Discharge Elimination System (NPDES) Stormwater Program	US Environmental Protection Agency (US EPA) (epa.gov/npdes)	This section of the US EPA website discusses how lifecycle stormwater management must be addressed during design so that construction water management (and ongoing rainfall after commissioning) is achieved within the necessary mitigation measures. Thoughtful design can allow for improved water retention and groundwater replenishment for the life of the facility.

Sustainable Packaging Design

These resources contain topical information and tools to explore, learn and challenge current practice–to move towards more sustainable packaging design. It's a diverse collection of websites, articles, videos, podcasts, and books that introduce sustainable design, recycling, circular economy, and sustainable packaging. Three are highlighted, with a more complete list at www.thecarbonalmanac.org/resources.

RESOURCE	DESCRIPTION
The Rise and Growing Importance of Sustainable Packaging Design | NS Packaging (nspackaging.com)	Article providing an overview of the movement toward sustainable packaging design.
Sustainability Guide: EcoDesign | European Regional Development Fund (sustainabilityguide.eu/ecodesign)	Primer on the concept and practices associated with eco-design packaging. The eight steps of the Ecodesign wheel – an initiative of the EcoDesign Circle – raise awareness on ecodesign among organizations and professionals.
Plastic Wars | Frontline Public Broadcasting Station (PBS) & National Public Radio (NPR) (pbs.org)	A documentary on recycling that takes an inside look at why and how the plastic industry created a narrative that would allow it to continue growing despite the negative impact its products were having on the environment.

Agriculture & Livestock Sustainability

These resources provide farmers (and others) topical information and tools to explore sustainability globally and locally. It's a diverse collection of sites and multimedia sources. Six are highlighted, with a complete list at www.thecarbonalmanac.org/resources.

RESOURCE	DESCRIPTION
Carbon Farming | Carbon Cycle Institute (carboncycle.org)	US-based organization working to join climate science and agriculture. Strategic partners include farmers, ranchers, researchers, public institutions, and businesses work to advance "natural, science-verified solutions that reduce atmospheric carbon while promoting environmental stewardship, social equity, and economic sustainability."
Research Institute of Organic Agriculture (FiBL) (fibl.org)	Switzerland-based global organization leading scientific and applied research on organic agriculture, with a special emphasis on transferring knowledge quickly from research to advisory work. Encouraging sensible development and training farmers on effective, sustainable practices around the world.
Global Agenda for Sustainable Livestock (GASL) (livestockdialogue.org)	A partnership of global livestock stakeholders from agriculture, government, education, and private sectors to build consensus toward sustainable food security and resource management while addressing equity, health, and growth challenges.
Regeneration International (regenerationinternational.org)	An international organization supporting the transition to regenerative agriculture through education, network-building, and policy work.
Natural Resources & Environment: Climate Change | US Department of Agriculture: Economic Research Service (USDA) (ers.usda.gov)	US Department of Agriculture web page that aggregates a number of articles, reports, and current statistics on climate change issues related to agriculture.
COMET Farm: COMET-Farm Tool | US Department of Agriculture (USDA) & Colorado State University (comet-farm.com)	COMET-Farm is a tool developed for farm and ranch carbon, and greenhouse gas accounting. The tool guides users through farm and ranch management practices including alternative future management scenarios and generates a report comparing current with potential future scenarios.

The Carbon Neutral Home

This section shows people how to make their homes more carbon neutral. Included are websites, articles, tools, videos, and podcasts addressing energy creation, storage, and usage. Includes information on products and tools to help people consume less poweror become energy independent ("go off the grid"). Four are highlighted, with a more complete list at www.thecarbonalmanac.org/resources.

RESOURCE	DESCRIPTION
Homeowner's Guide to Going Solar | Office of Energy Efficiency and Renewable Energy, US Department of Energy (US DOE) (energy.gov/eere/solar)	An extensive resource document structured as a Q & A with answers to questions like "How does solar work?" and "Is my home suitable for solar panels?"
Calculate Your Carbon Footprint | The Nature Conservancy (nature.org)	Use a calculator to estimate what one's current carbon footprint is. This can provide a basis to start from and an idea of what personal actions will make the most impact.
Net-Zero 101: The Secret of Building Super Energy-Efficient Homes (video) (greenenergyfutures.ca)	Green Energy Futures offers a documentary featuring net-zero home designing/building pioneers Peter Amerongen and Mike Turner.
The Eco Store (ecostoredirect.com)	Many brands and kits to compare for solar, wind, and water power systems, along with energy storage systems and 3D printers.

The Sustainable Consumer

These resources cover how to become a smarter consumer. From podcasts to articles to calculators, there's a wide range of information and tools available to help consumers make more sustainable choices. Four are highlighted to get you started, with a more complete list at www.thecarbonalmanac.org/resources.

RESOURCE	DESCRIPTION
Climate Change: How Consumers And Businesses Can Make A Difference | National Science and Media Museum (scienceandmediamuseum.org.uk)	The National Science and Media Museum in the UK brought together a panel of experts to address questions on how consumer behavior can make a difference to climate change and how businesses can make it easier to live sustainably.
Good Together Podcast, hosted by Laura Alexander Wittig and Liza Moiseeva (brightly.eco/podcast/)	Brightly co-founders Laura and Liza host a podcast for those "curious about a zero-waste lifestyle…(wanting) to know what the "circular economy" really means." Each 30-minute episode features daily, actionable tips to help you live your life more sustainably.
"Climate Change Food Calculator: What's Your Diet's Carbon Footprint?" | BBC News, Aug. 9, 2019 (bbc.com/news)	An interactive tool from the BBC to learn the carbon footprint of your foods.
"How to Reduce Your Carbon Footprint" | New York Times, Jan 31, 2019 (nytimes.com/guides)	This guide details the choices individuals can make to lessen their personal impact on the environment broken down into five sections: On the Road, in the Sky; On Your Plate; In Your Home; What You Buy; What You Do.

Education Resources About Climate and Sustainability

The Almanac team has compiled extensive resources for educators working in many different settings. Included are educators' guides—including one to accompany this Almanac—along with websites, videos, podcasts, books, and lesson plans. These resources are useful to educational institutions, community-based programs, and individuals of all ages. Five are highlighted to get you started, with a more complete list at www.thecarbonalmanac.org/resources.

RESOURCE	DESCRIPTION
Educator's Guide (thecarbonalmanac.org/177)	A document assembled by the Carbon Almanac Team to help educators lead conversations with their students, manage stress, and promote solution-based thinking. Educators' Guide references back to the Almanac for optimum effect.
Communicating Climate Change: A Guide for Educators, Anne K. Armstrong, Marianne E. Krasny, and Jonathon P. Schuldt. 2018.	Provides insight into how audiences engage with climate change information. Written for environmental educators. The authors are from Cornell University. Print and online open-access versions are available.
The United Nations Intergovernmental Panel on Climate Change (IPCC) (ipcc.ch)	This United Nations-sponsored body is assessing the science related to climate change. Comprehensive summaries of the latest research on climate change and its impact.
Guide - Talking to Kids about Climate Change: Top Tips to Explain Causes, Effects, and Solutions | OVO Energy guides (ovoenergy.com)	Explanations and a list of activities educators can use to help make the current climate situation real to kids without it being frightening.
Educator Guide - Talking to Young People About Climate Change | UNICEF & UNESCO (worldslargestlesson.globalgoals.org)	Free downloadable guide for educators. Content focuses on education, problem-solving, and hope. Targeted to ages 8-14.

Legal Resources for Sustainability

These resources include links to legal organizations that are fighting climate change. These include information for policymakers, advising clients of climate risks, model contractual clauses, and education materials for law schools and practicing attorneys. Four are highlighted to get you started, with a more complete list at www.thecarbonalmanac.org/resources.

RESOURCE	DESCRIPTION
Climate Change Legal Blog Archive (climatechangelegalblogarchive.com)	The Climate Change Legal Blog Archive is an aggregation of legal blog posts offering information, insight, and commentary on climate change law and climate litigation published by lawyers globally. The archive has legal climate change blogs, podcasts, and videos.
Earthjustice (earthjustice.org)	Earthjustice is a US-based nonprofit public interest organization litigating environmental issues. Serves public interest groups in communities impacted by climate change.
ClientEarth (clientearth.org)	ClientEarth is an international environmental law charity. They partner across borders, systems, and sectors. Their work focuses on changing the system through informing, implementing, and enforcing the law. Serves European citizens and NGOs in over 50 countries by advising decision-makers on policy and training legal and judicial professionals.
The Chancery Lane Project (chancerylaneproject.org)	The Chancery Lane Project (TCLP) is a collaborative effort of global legal professionals. TCLP works with lawyers to create contractual clauses ready to incorporate into commercial agreements to offer climate solutions.

Glossary

ORGANIZATIONS, MEETINGS, AND FRAMEWORKS

United Nations Framework Convention on Climate Change (UNFCCC) An international environmental treaty drafted at the Earth Summit in Rio de Janeiro, Brazil in 1992. This framework combats "dangerous human interference with the climate system."

Kyoto Protocol The first implementation of measures under the UNFCCC, adopted in 1997 by 192 state parties committed to reducing greenhouse gas emissions from 2005 to 2020.

Paris Agreement Adopted in 2015 by 196 state parties, it replaced the Kyoto Protocol commitments beginning in 2016. The agreement covers climate change mitigation, adaptation, and finance. It includes a "ratchet mechanism," where every five years parties are expected to improve their national pledges.

COP26 An international summit held in Glasgow, 2021, known as a Conference of the Parties (COP). This was the COP where the ratchet mechanism went into effect.

Intergovernmental Panel on Climate Change (IPCC) A United Nations body, established in 1988, that advances knowledge on human-induced climate change and evaluates its impacts.

Group of Twenty (G20) A group of 19 countries and the European Union that meet regularly to address issues such as the global economy, international financial stability, climate change mitigation, and sustainable development.

Activism Taking action in support of or against an issue.

Activist You.

Aerosol A suspension of fine solid or liquid particles in the air or within another gas. Aerosols can be natural or manmade.

America Used here to refer to people in the United States.

Anthropocene The era we are living in when manmade actions have significantly impacted the Earth and its environment.

Anthropogenic Any change that is manmade or caused by human activity.

Asteroid Minor planets of the inner solar system. A metaphor for climate change.

Billion A number equal to one followed by nine zeros. A thousand multiplied by one million.

Biodiversity Biological variability or variety. A measure of variation at a genetic, species, and ecosystem level.

Biomass An organic material (plant or animal) often used as fuel to produce heat or electricity.

Biofuel Fuels that are produced using contemporary processes from biomass rather than slow geological processes that create oil. These fuels can be used to replace gas, petrol, or diesel.

Carbon Budget The upper limit of global carbon dioxide (CO_2) emission that remains within a specific global average temperature.

Carbon Cycle The process through which carbon is exchanged between the biosphere, geosphere, hydrosphere, and atmosphere.

Carbon Capture and Utilization Absorbing or capturing carbon from the air and using it for other industrial purposes.

Carbon Dioxide Emission Equivalent [CO_2e] A measure of the impact greenhouse gases other than carbon dioxide cause to the planet. Defined relative to CO_2. In this book, CO_2 is often used to describe the overall impact of all greenhouse gases, and we use CO_2e when specifically describing the impact of other gases.

Carbon Dioxide Removal The process of extracting carbon dioxide from the atmosphere to be buried or stored for a long period of time. Also referred to as CCS (Carbon Capture and Storage).

Carbon Footprint The total greenhouse gas emission caused by an individual, organization, or country. Measured in CO_2e.

Carbon Neutral When the amount of carbon emissions produced is canceled out by the amount of carbon removed from the atmosphere.

Carbon Offset The removal of CO_2 or other greenhouse gases as compensation for emissions created elsewhere.

Carbon Sequestration The process of storing carbon in natural geological formations underground which traps the gas permanently.

Carbon Sink Forests, oceans, and other natural formations that have the capacity to absorb carbon dioxide from the atmosphere.

Clean Energy Energy generated via naturally replenished resources that do not produce CO_2 as a byproduct. Also known as Green Energy and Renewable Energy.

Climate The long-term pattern of weather in a region, typically over 30 years. Weather encompasses all conditions in the atmosphere (including temperature) at a specific location and time.

Climate Change Long-term shifts in temperature and weather patterns.

Climate Justice Addressing the ethical dimensions of climate change.

Climate Migration Due to changes in weather patterns (e.g., rising sea levels, frequent droughts, change in rainfall), people are forced to move away from a region where they've lived for generations.

Deforestation The removal of a forest or stand of trees to clear land for some other purpose.

Desertification A process of land degradation by which fertile soil becomes arid and biologically unproductive. The process results from drought, extreme heat, deforestation, or poor agricultural practices.

Dollar Used here to refer to US dollars.

Drawdown A projected milestone for reversing climate change in which greenhouse gases in the atmosphere are steadily declining.

Ecosystem An area in which living organisms and their physical environment interact through nutrient cycles and energy flows that sustain the whole system.

Electric Vehicle (EV) A vehicle either partially or fully powered by an electric source.

Emissions Greenhouse gases released through the combustion of fossil fuels and other human activities. A primary cause of increasing changes to the Earth's climate.

Emissions Trading A market-based system, sometimes called "cap-and-trade," designed to reduce greenhouse gas emissions by providing economic incentives to companies and countries.

Energy Efficiency Using less energy to accomplish the same task or provide the same product or service.

Erosion Natural forces like wind and water wearing away or moving soil, rock, and other earth materials from one location on the Earth's surface to another.

ESG Reporting The disclosure of an organization's data in the areas of environmental, social, and governance (ESG) impacts.

Fluorinated Gases Human-produced gases used in a range of industrial applications and manufacturing processes. The longest-lasting of all greenhouse gases, they remain in the atmosphere for centuries and contribute significantly to global warming.

Fossil Fuels Material containing hydrogen and carbon found underground. Produced over millions of years from decomposing plants and animals, and then extracted as coal, oil, or natural gas.

Geoengineering The intentional large-scale intervention in an environmental process that impacts the Earth's climate with the goal of halting or reversing climate change.

Geothermal Energy A type of renewable energy drawn from beneath the Earth's surface. Taps into heat that was generated in the creation of the planet and the radioactive decay of materials.

Gigaton (Gt) A unit of mass equal to one billion metric tons or 2.2 trillion pounds. A metric ton ("Mt") is 1000 kilograms. A gigaton of CO_2 is often expressed scientifically as "109 Mt CO_2."

Global Warming The gradual increase in temperature of the Earth's air, surface, and oceans due to human-caused increases in CO_2 and other greenhouse gas levels.

Go-giver Someone who shares *The Carbon Almanac* with a decision-maker.

Greenhouse Effect When the sun's radiant energy gets trapped by gases in the Earth's lower atmosphere and heats the planet's surface.

Greenhouse Gas (GHG) Tiny particles released into the Earth's atmosphere that absorb energy from the sun, preventing heat from leaving the atmosphere. Gases include carbon dioxide, methane, nitrous oxide, ozone, water vapor, and chlorofluorocarbons.

Hope Fuel for change.

Hydroelectric Power A renewable form of energy that produces electricity by using the power of moving water.

Ice Sheet A glacial ice mass greater than 50,000 km². About 99 percent of the freshwater on Earth is contained in ice sheets.

Indigenous Knowledge Local knowledge and behavior developed by communities over centuries, including how to best use natural resources.

Industrial Revolution The period in Europe and the US that marks the transition during the late 17th and early 18th centuries from an agrarian and handicraft society into one dominated by industrial and machine manufacturing.

LEED Widely accepted system organized by the US Green Building Council to rate building performance on sustainability and impact on the environment.

Megawatt (MW) Megawatt is a measurement for power, used to measure the output from a power source. 1 megawatt is one million watts of power. Gigawatt (GW) is one billion watts of power. Megawatt hour (MWh) is a unit of measure of electric energy equal to 1,000 kilowatt hours (KWh).

Methane (CH_4) A colorless, odorless, and flammable greenhouse gas, with 82 times the potential warming effect of carbon dioxide (CO_2) over a 20-year period. Scientific notation is CH_4.

Mitigation Reduction of something harmful.

Natural Gas A non-renewable fossil fuel. Used mainly for heating, electricity generation, and manufacture of plastics and other products.

Net-Zero Emissions Balancing greenhouse gas emissions caused by human activity with emission reductions.

Nitrous Oxide (N_2O) Also known as nitrous or laughing gas. A strong greenhouse gas with 300 times the heat-trapping power of carbon dioxide. Scientific notation is N_2O.

Nuclear Energy Power derived from energy that is released by splitting atoms at nuclear power plants in a process called fission. This term will also include nuclear fusion if it becomes practical.

Nuclear Fission A reaction in which energy is produced when an atom's nucleus splits into two or more smaller nuclei of nearly equal mass.

Nuclear Fusion A reaction in which energy is produced when two or more atomic nuclei of low atomic number combine to create a heavier nucleus.

Ocean Acidification The decrease of the pH of oceans caused by increased uptake of carbon dioxide from the atmosphere, leading to less healthy oceans.

Ocean Currents The continuous and directed movement of ocean water caused by several forces including wind, temperature, and salt level differences.

Organic Farming An agricultural system advocating environmentally friendly methods to grow crops or raise livestock, promoting ecological balance.

Ozone Layer A shield in the Earth's stratosphere that absorbs most of the Sun's ultraviolet radiation.

Peatlands Wetland ecosystems sometimes known as wetlands or swamps. The densest natural storehouses for carbon on land.

Permafrost Frozen land near the Arctic circle and high-altitude mountaintops.

Petrochemical A chemical product made from refined petroleum.

pH A scale indicating the level of acidity in a solution. Lower numbers are more acidic.

Photosynthesis The process that plants and other organisms use to convert light energy into chemical energy to fuel their growth.

Plastic Plastics are mostly made from fossil fuels in the form of oil, natural gas, or coal and always contain carbon and hydrogen.

Recycling Processing something, usually waste, so it can be used again in the same form or by creating new material.

Reforestation Planting seeds or young trees to replace a lost forest.

Runoff The portion of rain or snow on land that reaches streams, often containing dissolved or suspended material (e.g., pesticides).

Saltwater Intrusion Rising sea levels that bring more saltwater to low coastal land, harming soil and making water undrinkable.

Sea Ice Floating frozen seawater.

Sea Level Change Warming temperatures cause ice to melt and seawater to expand, leading to changes in sea level.

Soil Degradation Improper use of land that decreases the quality of soil.

Sustainability Producing in a way that does not use what can't be replaced and does not damage the environment.

Tidal Surge Rising seawater level caused by a storm.

Ton A unit for measuring weight. *The Carbon Almanac* uses the metric ton, equivalent to 1000 kg or about 2205 lbs. Sometimes spelled "tonne" to differentiate from the American "short ton" of 2000 pounds.

Waterborne Disease Microorganisms and toxic contaminants in water that cause illnesses and gastrointestinal problems. Often occurs after a severe rainfall from runoff.

Wetlands Land naturally saturated in water (e.g., marshes, mangroves, estuaries, bogs, lakes).

Wildfire An uncontrolled fire destroying an ecosystem.

Xebec A three-masted schooner, an early form of sustainable commercial transport.

🌐 **178**

🌐 CHECK OUR WORK

The Almanac is based on thousands of sources. Don't take our word for it. Look for this number at the end of an article and then visit www.thecarbonalmanac.org/999 (but replace 999 with your article number). **Dig deep and share what you learn.**

www.thecarbonalmanac.org

Find the sources for all the quotations and fact boxes at 🌐 888.

Acknowledgments

Without exaggeration, this book is the first of its kind. More than 300 contributors from 41 countries came together, all as volunteers, to build the entire Almanac. Along the way, the extraordinary Carbon Almanac Network team received support, good wishes, and optimism from some passionate, busy, and insightful friends as well. Special thanks to **Fiona McKean**, **Tobi Lutke**, **Michael Cader**, **Stuart Krichevsky**, **Pam Dorman**, **Adam Grant**, **Justin Brice Guariglia**, **Maya Lin**, **Shepard Fairey**, and **Kevin Foley** and the team at **Getty Images**. We could not have done this without the software and support of the Discourse team, including **Jeff Atwood** and **Sam Saffron**.

Also thanks to **Aaron Schleicher**, **Adam Umhoefer**, **Carla Vernon**, **Carrie Ellen Phillips**, **Dana Pappas**, **Debbie Millman**, **Dylan Schleicher**, **Geerhard Bolte**, **Katherine Shepler**, **Maddy Roth**, **Martijn Vinke**, **Michael Jantz**, **Michelle Kydd Lee**, **Rebecca Schwartz**, **Simon Sinek**, **Steve Pressfield**, **Tina Roth Eisenberg**, **Nathan Gray**, **Yukari Watanabe Scott**, **Iván X. Eskildsen**, and **Danielle M Fino**. Also, thanks to **Andrew Pershing**, **Ben Strauss**, **Daniel Gilford**, **Sam Miller**, **Chip Conley**, **Paul Hawken**, **Kevin Kelly**, **Stewart Brand** and the **Geoversity Foundation**.

Our font providers: **Kostas Bartsokas** at Foundry5 and ABC Dinamo.

The cartoons are from **Dan Piraro**, **Tom Toro**, and **Randall Munroe**. Generous ruckus makers.

And of course, thanks to **Niki Papadopoulos**, **Adrian Zackheim**, **David Drake** at Crown, and **Markus Dohle** and the wonderful team at **Penguin Random House**.

Thanks to **Ben Fry** / Fathom Information Design. Thanks to **Anders Hellberg** for the photo of Greta Thunberg. We're grateful to the Noun Project for the astonishing array of icons, and to **Scott Belsky** and his team as well.

Fact-checking: **Will Myers** and **Stevonie Ross**. Copy editing: **D. Olson Pook**. All errors are the responsibility of the creators; please visit thecarbonalmanac.org if you find an error, and we'll fix it. Indexing by **Lucie Haskins**.

Please check our sources at www.thecarbonalmanac.org. We are grateful for the significant body of work created by Our World in Data and other scientists and publishers online. Visit our site for direct links to each of the datasets incorporated in the Almanac.

Contributors

This group represents more than 40 countries, including Australia, Belgium, Benin, Brazil, Canada, Colombia, Costa Rica, Côte d'Ivoire, Croatia, Czech Republic, Denmark, Finland, France, Germany, Greece, India, Ireland, Israel, Italy, Jamaica, Kenya, Mexico, the Netherlands, New Zealand, Nigeria, Poland, Portugal, Romania, Scotland, Senegal, Serbia, Singapore, South Africa, South Australia, Spain, Sweden, Switzerland, United Arab Emirates, United Kingdom, United States, and Uruguay.

Aarón Blanco Tejedor	Bruce Clark	Donal Ruane
Abhishek Sharma	Bulama Yusuf	Dorothy Coletta
Adam Davidson	Cameron Palmer	Dr. Meenakshi Bhatt
Alberto Parmiggiani	Carlo Tortora	Elena-Madalina Florescu
Alessio Cuccu	Carlos Saborío Romero	Eva Forde
Alexandre Poulin	Casey von Neumann	Fabio Gambaro
Alexis Costello	Charlene Brown	Felice Della Gatta
Allyson Alli	Charles Dowdell	Fernando Laudares Camargos
Amy Maranowicz	Chirag Gupta	Gabriel Campbell
Andrea Hunter	Christopher G. Fox	Gabriel Salvadó
Andrea Martina Specchio	Christopher Houston	Gillian McAinsh
Andrea Morris	Colin Steele	Giorgia Lupi
Andrea Ramagli	Con Christeson	Helena Roth
Andrea Sakiyama Kennedy	Conor McCarthy	Hiten Rajgor
Andreas Andreopoulos	Corey Girard	Inbar Lee Hyams
Ángela Conde del Rey	Covington Doan	Inma J Lopez
Angelica Liberato	Craig Lewis	Isabelle Fries
Anna Cosentino	Crystal Andrushko	J. Thorn
Anne Marie Cruz	Dalit Shalom	Jasper Croome
Annie Parnell	David Kearns	Jay Wilson
Asante Tracey	David Kopans	Jayne Heggen
Ash Roy	David Meerman Scott	Jeff Goins
Azin Zohdi	David Olawumi	Jennifer Hole
Barbara Orsi	David Robinson	Jennifer Myers Chua
Barrett Brooks	Dawn Nizzi	Jennifer Simpson
Belinda Tobin	Debbie Cherry	Jennifer V Taylor
Benjamin Collins	Debbie Gonzalez	Jessica P. Schmid
Benjamin Goulet-Scott	Deepa Parekh	Jim Kennady
Blessing Abeng	Denis Oakley	Joaquin Ilzarbe
Boon Lim	Diane Osgood, Ph.D.	José Ignacio Conde
Brent Brooks	Dianne Dickerson	Kady Stoll
Brian Stacey	Dillon Smith	Kanakalakshmi Balasubramani

Karen Mullins

Kat Chung

Kate Shervais

Katharina Tolle

Kathryn Bodenham

Keary Shandler

Kelsey Longmoore

Kevin Caron

Kevin Lockhart

Kirsten Campbell

Kristin Hatcher

Kristy Sharrow

Kurt Hinkley

Lars Landberg

Laura Holder

Laura Shimili

Laurens Kraaijenbrink

Leah Granger

Leekei Tang

Leonardo Scopinho Heise

Lewis Thompson

Linda Westenberg

Lisa Blatt

Lisa Duncan

Lisa Oldridge

Lisa Sarasohn

Liz Cyarto

Lori Sullivan

Louise Karch

Lucy Piper

Luke Keating Hughes

Lynne E. Richards

Magdalena Zwolak

Maggie Hobbs

Manon Doran

Marcelo Lemos Dieguez

Margo Aaron

Marjolaine Blanc

Mark Belan

Mark Conlon

Mark Deutsch

Markus Amalthea Magnuson

Marty Martens

Maryanne Sherman

Massimiliano Freddi

Matthew Andreus Narca

Matthew NeJame

Maureen Price

Max Francis

"Maya" Aparajita Datta

Mayank Trivedi

Mel Sellick

Meredith Paige NeJame

Michael Bungay Stanier

Michel Porro

Michelle Miller

Michi Mathias

Monica Wilinski

Natalia Alvarez

Natasa Gacesa

Natashja Treveton

Nell Boyle

Nick Delgado

Noura Koné

Pasquale Benedetto

Paul McGowan

Philip Amortila

Polo Jimenez

Rachel Ilan Simpson

Ray Ong

Reginald Edward

Richie Biluan

Robert Gehorsam

Robert L. Hill

Roger R. Gustafson

Rohan Bhardwaj

Roma G Velasco

Ronald Zorrilla

Ryan Flahive

Sally Olarte

Sam Nay

Scott Ash

Scott Hamilton

Scott Papich

Sean Kim

Selena Ng

Seniorita Polyester

Seth Barnes

Seth Godin

Shaun McAnally

Sisi Recht

Stella Komninou Arakelian

Steve Wexler

Suparna Kalghatgi

Susan Hopkinson

Susan Z Martin

Susana Juárez

Sydney Alexandra Shoff

Szymon Kurek

Tania Marien

Teresa Reinalda

Tobias Kern

Tom Gelin

Tonya Downing

Tracey Ormerod

Virginia Shaw

Vivek Srinivasan

Winny Knust-Graichen

Yan Tougas

Yolanda del Rey Chapinal

Zrinka Zvonarevic

Find bios, photos, and more at thecarbonalmanac.org

Index